The Upshaws of County Line

An American Family

To Sherrie — remembering our long friendship

Richard Orton

THE UPSHAWS OF COUNTY LINE

AN AMERICAN FAMILY

PHOTOGRAPHS BY RICHARD S. ORTON

FOREWORD BY THAD SITTON
PREFACE BY ROY FLUKINGER

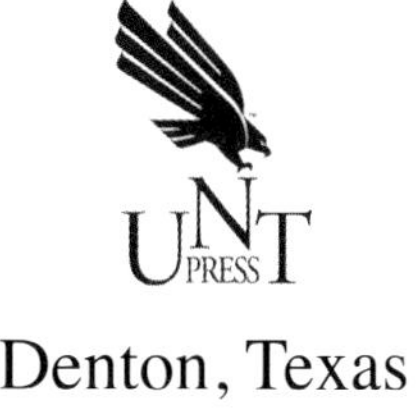

Denton, Texas

Printed in China

10 9 8 7 6 5 4 3 2 1

Permissions:
University of North Texas Press
1155 Union Circle #311336
Denton, TX 76203-5017

The paper used in this book meets the minimum requirements of the American National Standard for Permanence of Paper for Printed Library Materials, z39.48.1984. Binding materials have been chosen for durability.

Library of Congress Cataloging-in-Publication Data

Orton, Richard S., 1946- photographer, author.
The Upshaws of County Line : an American family / photographs by Richard S. Orton ; foreword by Thad Sitton ; introduction by Roy Flukinger. — First edition.
pages cm

ISBN 978-1-57441-571-1 (cloth : alk. paper) — ISBN 978-1-57441-584-1 (ebook)
1. Upshaw family—Pictorial works. 2. County Line (Nacogdoches County, Tex.)—Biography—Pictorial works. 3. County Line (Nacogdoches County, Tex.)—Social life and customs—Pictorial works. 4. African American farmers—Texas—County Line (Nacogdoches County)—Pictorial works. 5. Agricultural colonies—Texas—County Line (Nacogdoches County)—Pictorial works. 6. Freedmen—Texas—County Line (Nacogdoches County)—Pictorial works. I. Sitton, Thad, 1941- writer of preface. II. Flukinger, Roy, 1947- writer of introduction. III. Title.
F394.C8125O78 2014
976.4′182—dc23
2014020875

The Electronic edition of this book was made possible by the support of the Vick Family Foundation

Contents

Foreword

By Thad Sitton

Freedmen's settlements like County Line were independent communities of African-American landowners and land squatters that formed in the eastern half of Texas during the years after Emancipation in 1865. Similar places took root in other former slave states. These "freedom colonies," as black people sometimes called them, were anomalies in a post-war South where whites quickly resumed social, economic, and political control, and the agricultural system of sharecropping came to dominate.

Freedmen's strong desires for land, autonomy, and isolation from whites motivated formation of these independent black communities. After the 1865 rumor that the federal government soon would provide all ex-slaves with "forty acres and a mule" proved false, most freed persons remained in the countryside and took employments with white landowners as day laborers, sharecroppers, or share tenants. Another large group of ex-slaves moved to segregated quarters adjacent to white towns. But a minority of former bondsmen set out to get their forty acres and a mule quite on their own, and a good many of them succeeded.

It was lonely out there in the white countryside, so a lot of these people formed dispersed farming communities—"settlements," Southerners often called them, whether blacks or whites resided there—places un-platted and unincorporated, individually unified only by church and school and residents' belief that a community existed.

Up in the sand hills, down in the creek and river bottoms, and along county lines, several hundred Texas freedmen's settlements came into being between the late 1860s and 1890. Many established themselves on pockets of wilderness, cheap land, or neglected land previously little used for cotton agriculture. Unimproved land was abundant and almost worthless in 1865, especially if covered with big trees.

One of these was the community that came to be known as Upshaw or County Line. Sometime around 1870, three Upshaw brothers—Guss, Jim, and Felix—moved to a wilderness area along the Angelina River bottoms in western Nacogdoches County and started an independent settlement. The brothers had been slaves of someone living in or around Douglass, a small market town in southwestern Nacogdoches County. Perhaps they were squatters with someone's permission, perhaps they squatted in a wilderness area claimed by no one, but in any case their community existed for years before residents began to brave the white courthouse to get formal titles to their lands.

Many enslaved people knew only farming and had wanted their own land after Emancipation, seeing it as the only real escape from white domination. Land bestowed a kind of concrete freedom, a daily practical reality of being at liberty replaceable by nothing else. A black representative in South Carolina said of his fellow freedmen, "Night and day they dream of owning their own land—it is their all in all."

Dreaming was all well and good, but it took a lot of nerve to go out into the wilderness, surrounded by potentially hostile whites, and try to get land for yourself. The three Upshaw brothers did that and so joined the ranks of other legendary founders like "Bear Dave" Henry of the Pelham settlement, Navarro County, who lived in three centuries (1790–1905) and once killed a bear with his own hands. When you work back through oral history and as-told-to accounts of the origins of these settlements to pick up the faint trail of the founders, you sense other Bear Daves, willing to face black bears, the white-robed KKK, and a crushing poverty. Some of the pride felt by latter-day residents of County Line, which may be seen in Richard Orton's portraits, derives from a sense of giants in the earth.

My tenant-farming great-grandfather, Jim Ed Harrell, lived here and there on different places in the Douglass hinterlands during the 1880s and 1890s, and he doubtless knew of the legendary Upshaw brothers of County Line. Felix was a competent farmer, but Guss and Jim had reputations as mechanical wizards not dissimilar to Harrell's own. Jim Upshaw the blacksmith ran a combined steam gin, sawmill, and grist mill, and Guss Upshaw was a master blacksmith and craftsmen who could, as people once said of such men, "fix anything but a broken heart." Guss built wagon wheels from scratch and could fabricate and replace every metal, wooden, and leather part needed to repair a wagon or a plow. After a quick look at another man's, he constructed a fish weir across the nearby Angelina River and supported his family on fish sold from it. People said Guss made split oak baskets so finely crafted and tightly woven that they held water.

Who knows what Jim Ed Harrell thought about the Upshaws, but local whites often regarded the residents of nearby freedmen's settlements like County Line with a variable mix of increased respect (even grudging admiration), caution, and active disapproval. Some whites believed that blacks were not even supposed to own land. Settlement people were different than blacks on the share-cropper farm or in the town quarter, and especially so when they were on home ground. Local whites knew that, but outsiders entering such communities might get a shock.

Although they might discreetly refrain from painting their houses and might assume the proper dress of field hands when they went in to town, whites often suspected that settlement folks were "uppity," and from time to time something like that did show through. A county away to the west, an old man in the Wheeler Springs settlement liked to play cow-horn revielles to his neighbors, black and white, for miles around, and whether they liked to hear them or not. And in the early 20th century a son of Guss named N.E. (his full given name) felt free to exchange threats with certain white stockmen who resented his numerous hogs and cattle competing with their animals on the local free range. In 1938 Elelia Upshaw, wife of yet another son of Guss, dared to fire her shotgun at two trespassing whites trying to compel a forced purchase of her calf.

This was indeed uppity. Counter threats and assertions of individuality like the blowing-horn concert and the violent stock defense were not the norm for blacks at the height of Jim Crow segregation, where white-dominated communities each had their special sets of rules for blacks moving in and around them—rules backed up by the threat and reality of violent assault. Blacks often didn't dare to defend themselves.

Freedmen's settlements like County Line can be fully understood only in the context of that Jim Crow world, which was so bad that many people, whites and sometimes blacks, have tried to forget it, or at least the worst of it. Jim Crow tasted bitter, and the taste grew tiresome and old. African-Americans had to dress a certain way, move down certain streets on their routes to the courthouses or mercantile stores, and obey certain curfews—or else. The customary rules of segregation were both complicated and different from place to place, and the most trivial violation might escalate to violence.

One county to the south, a black man forgot and wore his hat into an Apple Springs store, and the owner shot him to death. Simple assaults and killings like this one far outnumbered the rare public lynching, and occasionally whole communities were attacked. Forty miles to the west of County Line on the Houston—Anderson County line the freedmen's settlement of Slocum came under murderous assault by a white mob in 1910. That was Guss, Jim, and Felix Upshaw's dangerous world and the one their children lived most of their lives in.

The nearby county seat towns of Nacogdoches and Rusk had especially obnoxious sets of such rules and customs, most rural black people believed, and white rules also governed black behavior in smaller towns and most of the countryside. But not in County Line, Winter's Hill (twenty miles away), or other freedmen's settlements. In those places, Jim Crow's writ, such as it was, did not run, and blacks behaved like they were on their ground.

And they were, they stood on their own storied earth, their graveyards full of legends. At the freedmen's settlements, independence and isolation exacted a certain price, but residents chose to pay it. Children there grew up with their self-esteem intact, protected from the "death of a thousand cuts" inflicted on black people by the laws and customs of Jim Crow in the white-dominated towns and countryside outside.

This made a difference, and it showed up in the high-achievers that came from freedmen's settlements. Oprah Winfrey spent the first years of her Mississippi childhood in one of them. A treasurer of the United States (Azie Taylor Morton) came from St. Johns Colony, Caldwell County, and a National Merit Scholar and the first black student to enter Rice University came from Center Grove, Houston County, among numerous other examples. It is no accident that black folklorist John Mason Brewer, J. Frank Dobie's friend and colleague and the first African-American scholar to be entirely assimilated to the state's scholarly establishment, had grown to manhood at Cologne, Goliad County.

It also showed up in reluctance to join the general exodus from the countryside that began during the Depression 1930s and continued into the 1950s. Many people at freedmen's settlements have refused to entirely sever their links to the land, and others have returned.

When Richard Orton began to visit County Line in 1988, Monel and Leota Upshaw and a hard core of older residents still resided on their hallowed ground, and their children and grandchildren returned for visits several times a year, even the group living in distant Oakland, California. Some of these people plotted a permanent return. Marion Upshaw, grandson of freedman Guss Upshaw, purchased seventy acres once part of great-uncle Jim Upshaw's land. He bought this property for the clear spring that once served as County Line's drinking water, the foundations of Uncle Jim's steam mill, and the small log structure still showing the marks of Jim Upshaw's broad ax. Upshaw hoped that his children would bond with this land he felt so strongly about and where he felt the presence of his dead father and mother every time he visited.

There are giants in the earth and other "presences" at County Line, and when you look closely at the photos of Richard Orton perhaps you can sense some of them.

—Thad Sitton, Austin, Texas

Preface: Come On In

By Roy Flukinger

I have confidence in the laws of morals as of botany. I have planted maize in my field every June for seventeen years and I never knew it come up strychnine. My parsley, beet, turnip, carrot, buck-thorn, chestnut, acorn, are as sure. I believe that justice produces justice and injustice injustice.

—Ralph Waldo Emerson

When Russell Lee established a photography program in the Department of Art and Art History at The University of Texas at Austin in the 1950s, he consciously elected to not turn out a generation of photojournalists in imitation of himself. From the first Lee set out to challenge his students, both undergrads and grads, to explore their own expressive interests and critical questioning of the art form. Lee would always encourage the personal investigation of the medium's potential—he was too much of a humanist not to do otherwise—but it was always done with respect to the imagery and ideas that were generated by those in his classes.

Lee would succeed in this endeavor not by showing his own work but rather by having his class members show theirs instead, while discussing everything from the nuts and bolts of technique to the expansive evaluation of artistry, design and motivation. A critique by this photographer-mentor would be unsparing but never harsh, and insightful without lapsing into the didactic. Lee always encouraged his potential artists to "go out and look," and then to return with fresh and new perspectives on their art. Only then, when debate and discourse had turned portfolio viewing into a shared learning opportunity, might Russell Lee the famous photographer share some anecdote or observation from his vast life experience.

It was from his own history that Russ would draw upon one or more secrets-that-were-really-no-secrets-at-all. When asked how, during his long journeys across '30s America photographing for the Farm Security Administration, he was able to gain the trust of so many strangers who came before his lens, he revealed that it all stemmed from his overwhelming interest in people. The simple procedure was that, whenever he arrived in a new town or farming community, he would initially leave the camera behind in the car. On his first day there he would go around, meet the people, learn their names, tell them who he was and what he was doing there, learn what interested them, and ask for their suggestions about what might make for interesting photographs. Simple, direct, obvious—and straight from the heart. Russ said that he could always tell that he had succeeded when, usually by the end of the first day, most people were asking him, "Well, if you're a government photographer, where's your dang camera?"

The light is what guides you home; the warmth is what keeps you there.

—Ellie Rodriguez

I think of Russ's lesson here—although he would likely term it more of an anecdote—whenever I consider the body of imagery that Richard Orton has produced over the years that he has photographed in and around County Line, Texas. Both of the men remind me of each other—plain, direct, soft-spoken and sincere in both their intent and their vision. They share in the firm belief first expressed by the philosopher-farmer Ralph Waldo Emerson that "The production of a work of art throws a light upon humanity." Both were impelled by the human stories that they believed could be depicted, revealed and ultimately honored through their cameras. And both took the time and the important first step to seek permission to enter into the lives of others.

As Orton plainly states: "I was curious." That is a key factor to remember in this body of work. His primary motivation to leave his comfort zone, meet new and different folks, and bring his camera along, arose from a human interest stemming from both the past fact that there existed in this corner of Texas a historical community founded by former slaves and by the present-day fact that a generation of descendants still called this place their home. Too often the impetus to create a picture story about others is generated by more complex incentives: the nosiness of the newshound, the indifference of the sociologist, or the prejudice of the politically motivated.

I have known Orton for years and can say with certainty that such obtuse motivations do not lie behind his photography. His sense of fairness stems from his Texas heritage and his years of public service. Like the Upshaws he recognizes that country time is not city time. He shares their affinity to the smaller towns and rural communities and family farms that are still the honored foundation of this state. I do not doubt that it only took a moment for Monel Upshaw to recognize in Richard the face of another honest man who would have no truck with exploitation or betrayal. A promise was made, a handshake was given, and for nearly half a generation now a photographer and his camera have never betrayed that trust.

I knew what slant of light would make you turn over. It was then I felt the highways slide out of my hands. I remembered the old men in the west side cafe, dealing dominoes like magical charms.

—Naomi Shihab Nye

In a classic sense Orton's photographs have held to the primary tenets of documentary photography. While every photographer is to some degree subjective, a sense of commitment to a certain faithfulness can always be maintained. Orton is not an Upshaw—he knows it and they know it—but he has been privileged to share in their lives, their living spaces, their memories and their beliefs. And that is a shared experience that all of the participants have taken to heart.

One is especially impressed by the diversity of Orton's imagery. There are interior and exteriors depicted. Strong portraits, both singular and in groups, can reveal both character and dreams. Pictures of events range from the monumental to the prosaic. Scenes in the family spaces of homes contrast with those within the community spaces like the church and the general store. Details of faces or bodily gestures are juxtaposed with larger scenes of a boisterous family gathering like their August Meet or the potent grief of a loved one's funeral. And I thank him for permitting us the modest opportunity to celebrate the rich intimacy that he discovers in the simple magic of reading or the elegant democracy of a domino game.

In depicting the world of the Upshaws—and in providing us with the rich counterpoint of their own words throughout the text—Orton does not ignore the reality of their world. The family knows the pinch of hard economic times and some of the vicissitudes of class and racial prejudice. But these are not the "heavyhearted pictures" (as Janet Malcolm characterizes them) that another FSA photographer, Walker Evans brought back from his and James Agee's times with three farm families in 1936 Alabama. In their particular fashion, and especially with the shared strength of their family, the Upshaws depicted here continue to endure. And the problems that most concern them today—preserving the family, keeping the land and the homestead, worrying about the future of the next generation—are by and large still to be found among most American families.

But maybe it is best not to compose a hymn
or chisel into tablets the code of his behavior
or convene a tribunal of men in robes to explain his words.

Let us not press the gold leaf of his name
onto a page of vellum or hang his image from a nail.
Better to fly over this little town with nothing
but the hope that someone visits his grave

once a year, pushing open the low iron gate
then making her way toward him
through the rows of the others
before bending to prop up some
flowers before the stone.

—Billy Collins, "Flying Over West Texas at Christmas"

As with many significant photodocumentary works, Richard Orton's coverage of County Line is one of multiple facets—bracketed by the arc of Monel's final years and passing, and embellished with imagery of the daily life on the homestead and the comings and goings of many family members. Added to this are the stories of the Upshaw women and men that Orton has recorded—voices of transition and change, memories of experience and age, reflections upon youth, and profound expressions of the strength and hope that are reflected in the photographs.

One of photography's most original sages, Minor White, once mused upon the fact that the true success of portraiture depended upon the element of "mutual trust and mutual vulnerability" as they were shared between photographer and subject. It has always seemed to be even much more expansive and profound—in that, beyond the mere process of recording the face, the experience extends to the multiple levels of human interaction and to all the dimensions of photography that can encompass them as well. There is much to commend those photographers whose certainty of vision has consistently embraced a faith in the very human.

What brought Richard to take that dusty road over to County Line may have, indeed, been a curiosity in a different people and a different racial culture in Texas other than his own—although knowing the man enough I can state unequivocally that the quest was also impelled by his very heart as well. Regardless of sociological interpretation or studied cultural resonance, what is most important is that he has, through his own insightful imagery and the words of the Upshaws themselves, brought us another homecoming, found us another branch of our American family, and given us an August Meet that is both deeply personal and profoundly universal. I am grateful that his photographs share with us all the traditions of home and church and land that this family has so generously shared with him.

As Butch cleared any obstacles out of the river so the Upshaw family kids could "Come on in!"—so too does Richard call out to us all. It is an imprecation that Russell Lee would encourage and an invitation for all of us to share.

Acknowledgments

I dedicate this book to my parents, Sidney William and Daisy Gribble Orton, who were willing to share me with the Upshaws, and to Edward Monel and Leota Freeman Upshaw, who welcomed me in their home and made me a member of their family.

I thank Dr. F. E. "Ab" Abernethy for making me aware of County Line, and for introducing me to Marion Upshaw, who I thank for inviting me to County Line in the beginning.

Ouida Dean had researched and written a paper on the history of County Line before I arrived which was very helpful to me in the early days.

Bill Kennedy introduced me to photography in the early 1980s and helped me build a darkroom in my house, enabling me to learn the craft and make photography my creative outlet.

Karen DeVinney, my editor at the University of North Texas Press, has been a pleasure to work with, being a novice in the world of book publishing myself.

Beyond that there are too many to name . . . all of Monel and Leota's family who I came to know, immediate and extended . . . other folks and families with homes in County Line.

Going to County Line put me in a different frame of reference, one that initially introduced me to a different cultural environment, but in the long run simply reframed and broadened my appreciation for humankind.

I offer a special thanks to those listed below. Their substantial monetary contributions helped make the publication of this book possible . . .

Madge Gribble Askonas
John Anderson
Claude and Annie Upshaw, Sr.
Michelle Ridlehoover
The East Texas Historical Association

Richard S. Orton
April 15, 2014

Going to County Line

By Richard S. Orton

I went to County Line for the first time the day after Thanksgiving in 1988. Marion Upshaw invited me out to meet his parents. I had just discovered that former slaves founded their own communities in East Texas shortly after the Emancipation and that County Line was such a place. I had no idea that former slaves could own their own land ten or so years after being freed. I was curious . . . and amazed! How could African Americans a decade or so out of slavery acquire hundreds of acres of land and start their own autonomous communities?! No history book I had read prepared me for this reality. I was curious about what effect relative independence from white authority had had on the generations born and raised in County Line.

County Line is also known as the Upshaw Community because three brothers—Guss, Felix, and Jim Upshaw—were the first to go there, and they started the community. Other families moved in shortly afterward. They called it County Line because it was on the Angelina River, the line between Nacogdoches and Cherokee Counties in East Texas.

Marion Upshaw was born and raised in County Line. When we met I told him that I wanted to learn about the history of the community and make photographs there. He invited me out, as I said, to meet his parents. His father had apparently had a bad experience with some other white folks who wanted to photograph the place, so that didn't work. Marion suggested that I get back in my truck and go a bit further down the road and introduce myself to his Uncle Monel.

I remember asking myself at that point why I would expect anybody in County Line to want to have anything to do with me. The drive down to Uncle Monel's place took only two or three minutes, but it seemed much longer. I was out of my comfort zone and not sure what to expect.

I parked my truck under the big red oak in front of the house, got out, and saw a large man I took to be Monel standing in a hog pen thirty or forty yards away. I walked over, introduced myself, told him what I wanted to do . . . and . . . would that be alright with him?

He paused for a moment . . . then said, "Well, I don't see why not."

That was all there was to it. From that moment forward he and his wife, Leota Freeman Upshaw, made me welcome in their home. As I write this, twenty-four years later, I am still amazed at their openness and generosity. But, as their children have told me repeatedly, that's how "Butch" and "Otie" were.

On the other hand, some of their children needed a little more convincing. When I went to my first Homecoming the following August I was met at the gate by Alfred "Tenchie" Upshaw, one of their thirteen children and a California Highway Patrolman. He very politely, and very firmly, asked me a number of questions designed to determine just exactly what I was up to. I was not offended. He had every right to check me out and this was his first opportunity. We became fast friends until his untimely death in 2006. Tenchie wanted to move back to County Line so badly that it became a joke with some of his California friends and family. As was the case with his brother, Claude "Bubba" Upshaw, Tenchie was more interested in moving back to the country than either of their wives, who were not country folk. But before his death, Tenchie began the process of creating a home in County Line for himself and anyone else he could bring along. He bought an old house and, at some expense, had it moved to County Line on narrow county roads. He died shortly thereafter, but his widow and stepsons vowed to complete the project of making a good, livable house out of it, and they have.

Tenchie's desire to return to County Line was typical of those of his generation who grew up there. They know where "home" is, and many who went elsewhere to get educated and make a living have come back or plan to do so.

My interest in County Line grew out of curiosity about the existence of a relatively independent community of African Americans in East Texas so soon after slavery and the effect of that autonomy on succeeding generations. Though I had been born in Nacogdoches, I knew nothing about the mostly independent black communities spread throughout East Texas established at about the same time as County Line.

For the first two or three years I was self-conscious during my visits to County Line. I was almost always the only white person there. It was a little like my two years in the United States Peace Corps in Liberia, West Africa, where I had the experience of being a cultural minority. County Line, too, had a different frame of reference, one defined by rural, African-American sensibilities and values with minimal attention to white culture. It took a while for me to "acculturate" within these surroundings.

To be viable, any community must have a cultural center. County Line had a school and church that provided that center. After desegregation in 1968 the school was closed and the church was all that was left. It became clear to me that if I wanted to know the community I needed to visit the County Line Baptist Church.

Reverend Wickware pastored the church when I first entered it in 1989. Because I was brought up in the Episcopal Church, the County Line Baptist Church was a very different experience for me, more informal than any white church I had attended . . . and more emotionally "participatory."

Reverend Wickware was old school. At the climax of his sermon he would begin speaking in a chant-like way, walking in a deliberate fashion back and forth behind the lectern. The folks in the pews responded… amen… say it preacher!

I learned a lot about County Line in that church. The community was most present there: in church, it seemed to me. I came to understand that for a long time the County Line Baptist Church had been the center of the community, not just in a religious or spiritual way, but in a far more fundamental way that stimulated strong relationships and helped them to survive as a community in the old days.

It was there in the County Line Baptist Church that Leota Upshaw, in 1992 on Homecoming Sunday, claimed me as her foster son . . . as they say, "in front of God and everybody." I went from being an only child to having thirteen brothers and sisters in the time it took to say "amen." (Though, at the time, the enthusiasm of the thirteen for this development was less than certain.)

However surprising this was to all of us then, I have been made to feel very much a part of the family in the ensuing years. The Upshaws have changed my life. I am a more complete person because of them.

Many of the photographs in this book were made during the annual Homecoming occurring on the weekend of the second Sunday in August. Known in the old days as the "August Meet," the second Sunday in August was and is an occasion for people who have left the community to return and renew relationships with family and friends. People from neighboring communities come, too, and the Sunday evening church service is the beginning of a weeklong revival.

More recently, the Saturday before Homecoming has become a reunion day for each of the individual families in the community, with plenty of food and socializing outside under the trees. The afternoon and evening are filled with card games, dominoes, and lots of talk.

At its height in the mid 20th century County Line included dwellings for about twenty families. The fields between their houses were filled with crops. Neighbors' houses were clearly visible across the fields. Nowadays those fields are filled with brush and pine trees. Those who remember the earlier days miss the clear views across the community. The trees are signs of a community in decline for them.

As Monel and Leota's children grew up and finished school, they, and others of their generation, left the community to pursue employment and education elsewhere. The size of County Line diminished accordingly. When I arrived in 1988 there were only four or five family homes left. Since then, though, several who left years before have returned. Others have created dwellings to which they return whenever they can.

The open question is this: What will happen to County Line when the generation of Monel and Leota's children and their first cousins is no longer around?

I hope that this book conveys some sense of the experience of growing up in County Line and of the significance of such communities in our common history.

Photographs

The Upshaws of County Line: An American Family

The fact that our ancestors were able to establish and build a church fresh out of the bonds of slavery and that we are able to maintain and improve it is a tribute to them, not to us. We are what we are because of what they were made of.

—Beatrice Upshaw

County Line is not a utopian place where all the little Negroes came together and built this community. That's not it. It was struggle the whole time living there in coming to terms with who I was as a person. It's a beautiful place, it's a lovely place, but it is a place of struggle. Why would anyone want to live there? [The founders] carved this place, this culture, this history out of something that no one else really saw. We made it our own. The people before me made it their own.

—Elia Upshaw Ali

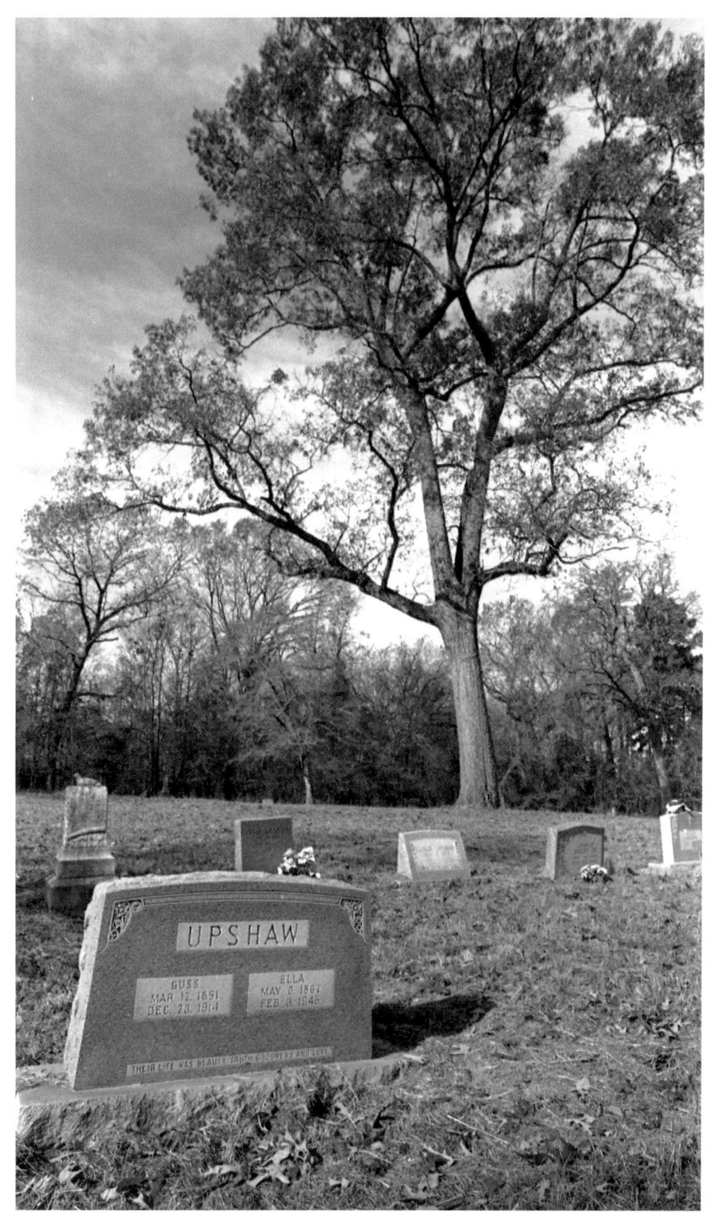

Stonewall Cemetery (1990)

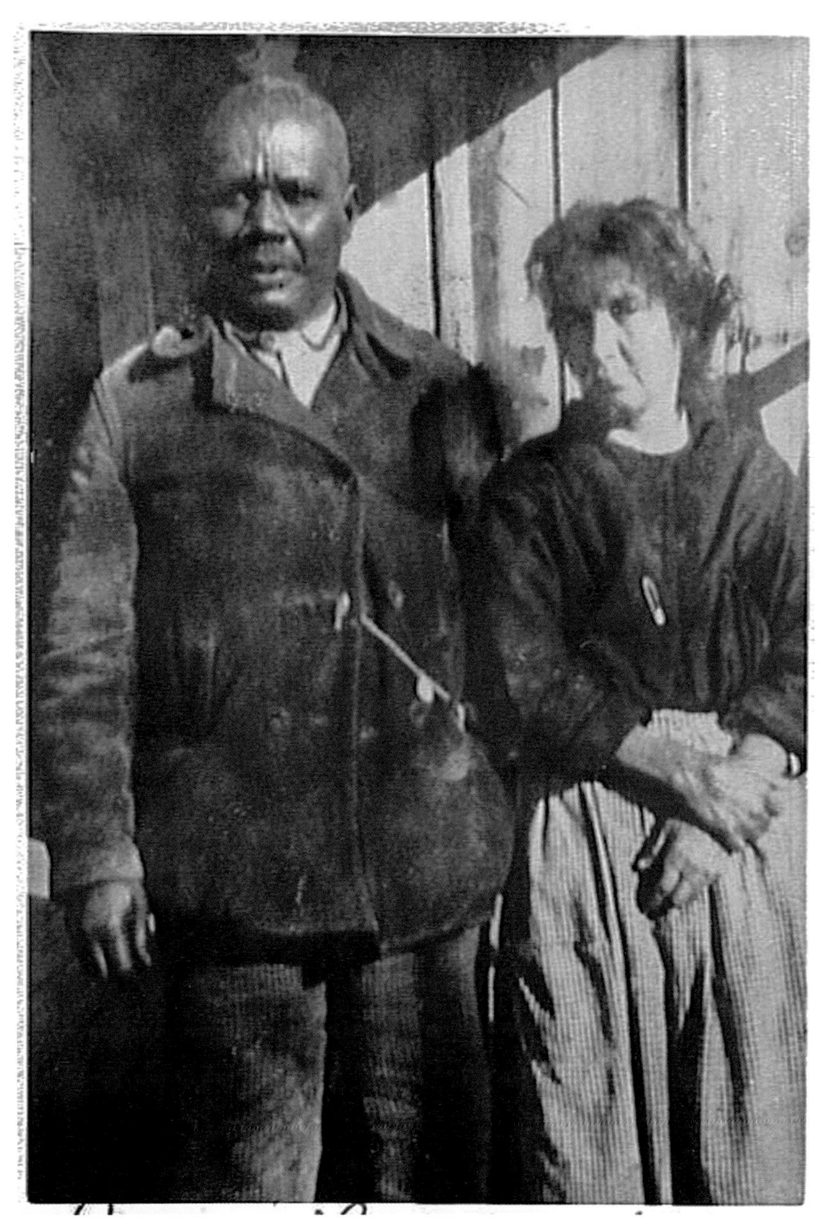

James and Laura Upshaw c. 1900

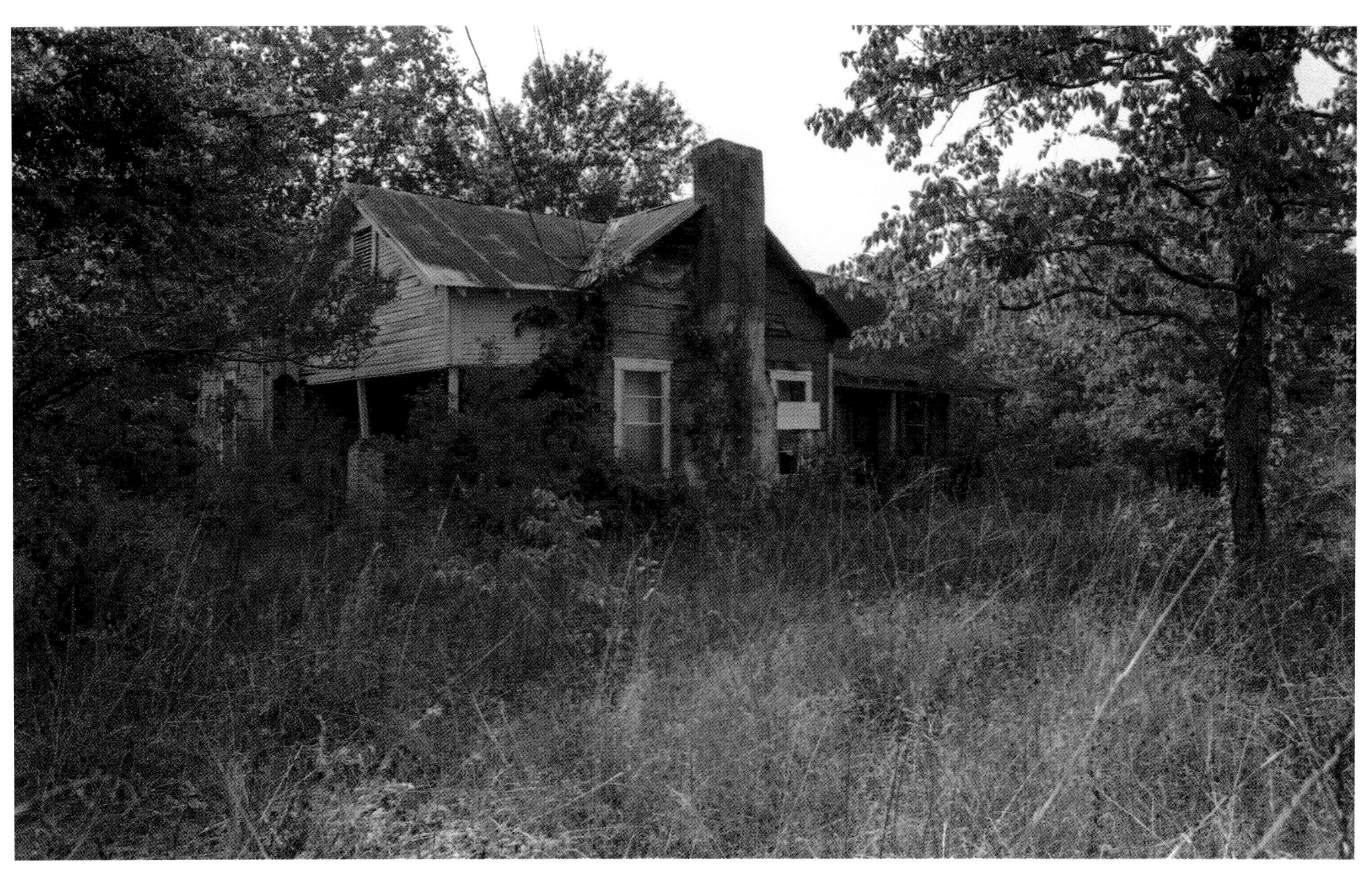

Home of Jim Upshaw (1998)

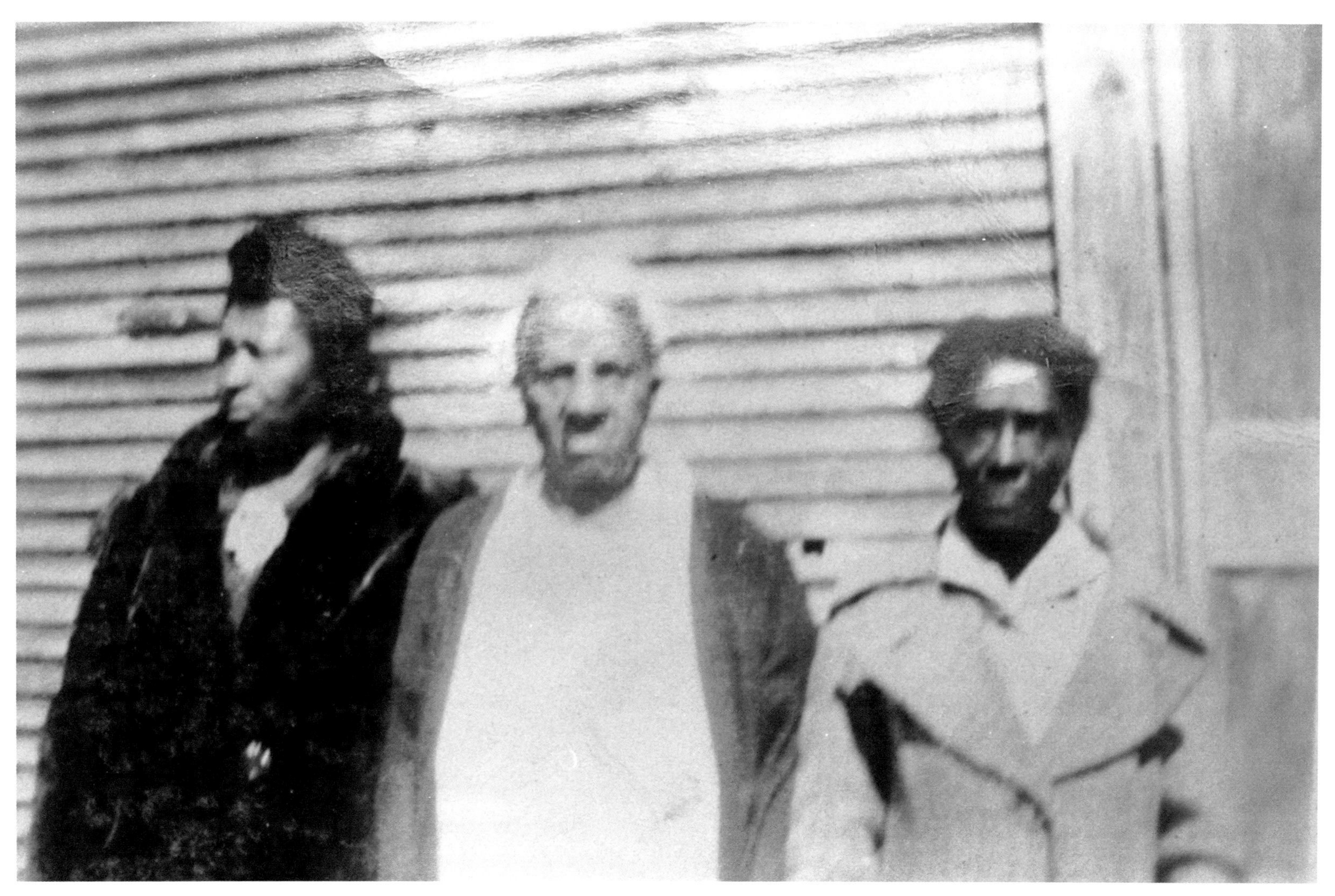

Unknown Woman, Ella Mathews Upshaw, "Little" Laura Upshaw (c.1930)

I did not appreciate growing up as part of a huge family. Did not appreciate all the "nots," and the "not haves" that go along with being part of a huge family. But as I grew older I certainly did appreciate that sense of togetherness that was a part of our growing up as kids. So that's one thing that I'm really proud of.

—Maye Upshaw Ham

It was a joy growing up with my mother and my stepfather. My mother taught us values, and she taught us to get along. She did not allow us to fight, and she didn't even want us to argue. If you started fighting you'd get a whipping. She instilled that in us, and to this day we have harmony in our family. That's because of the way they raised us.

—DeMorris Skinner Young

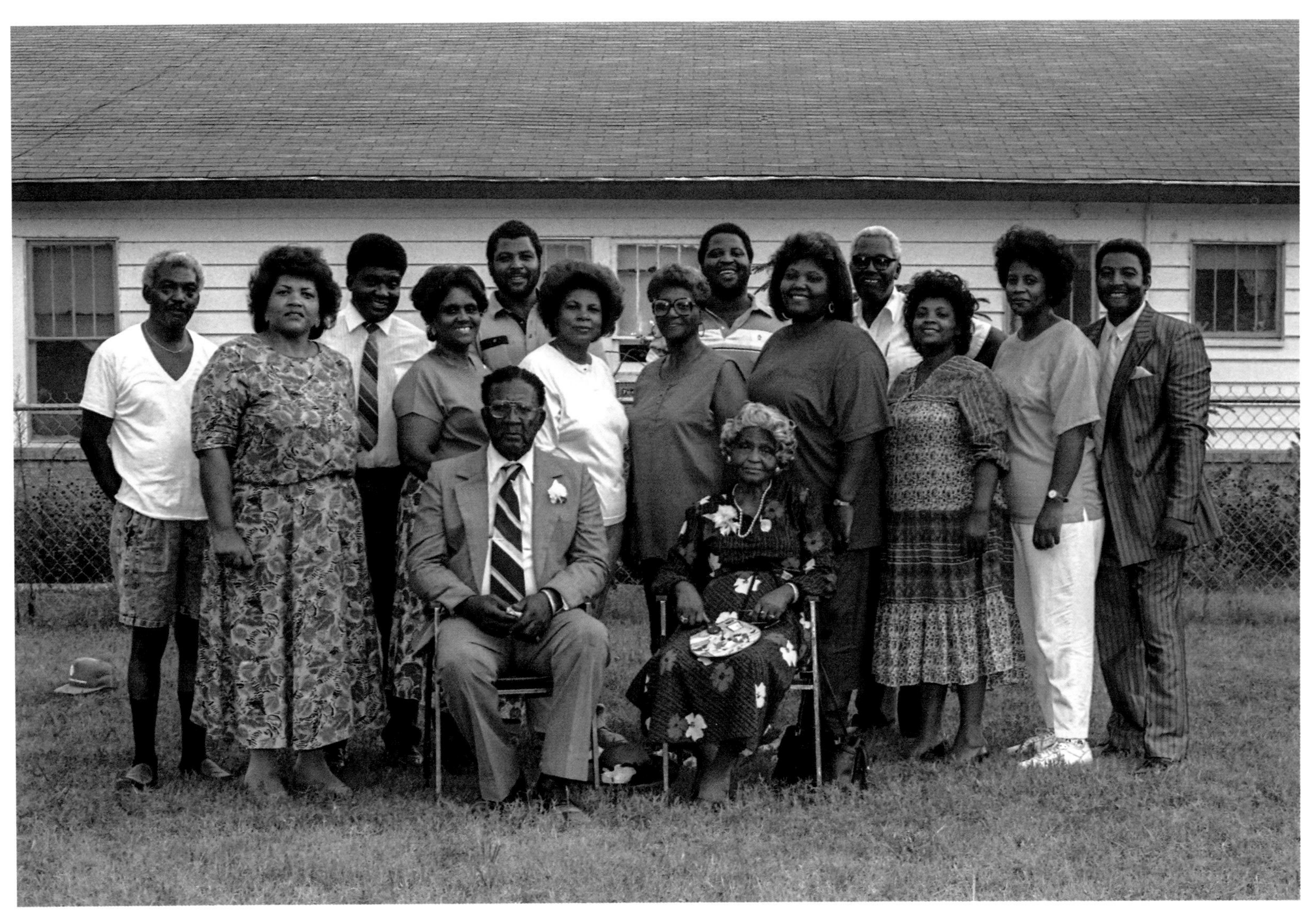

The Monel and Leota Upshaw Family (1990)

My daddy was a tree. As stout, massive and dignified as the red oak. He was: Unshakeable, Unmovable, and Unchangeable. A provider, a protector, a pillar, a philosopher. He shielded us kids from physical dangers and from certain other harsher realities of life—realities we would surely experience as adulthood and its accompanying independence came upon us.

We lived on a small farm a mile or so east of the Angelina River. On hot summer days our dad would load us onto the back of his pickup truck and drive us down to the river. Once there, he would get out of the truck and say, "Now you young 'uns stay in the truck 'til I tell you to get off." He'd begin to wade out into the shallows, splashing noisily and yelling at the top of his lungs. He would do this for five or six minutes, an eternity to us kids left sweating and wriggling in the truck's bed, anxious to feel the cool water on our hot, dusty bodies.

Finally. Finally! He would yell, "Come on in!" Off the truck we would jump, girls giggling, boys grinning, excited at the prospect of frolicking in the water. We would wade in to the river and cavort happily under the watchful eye of "Butch."

Not until I was almost an adult did I realize why Butch went into the water first. He did this to scare away the poisonous water moccasins and dangerous snapping turtles so common in those waters, protecting us from dangers. Dangers to which we were blissfully oblivious.

—Maye Upshaw Ham

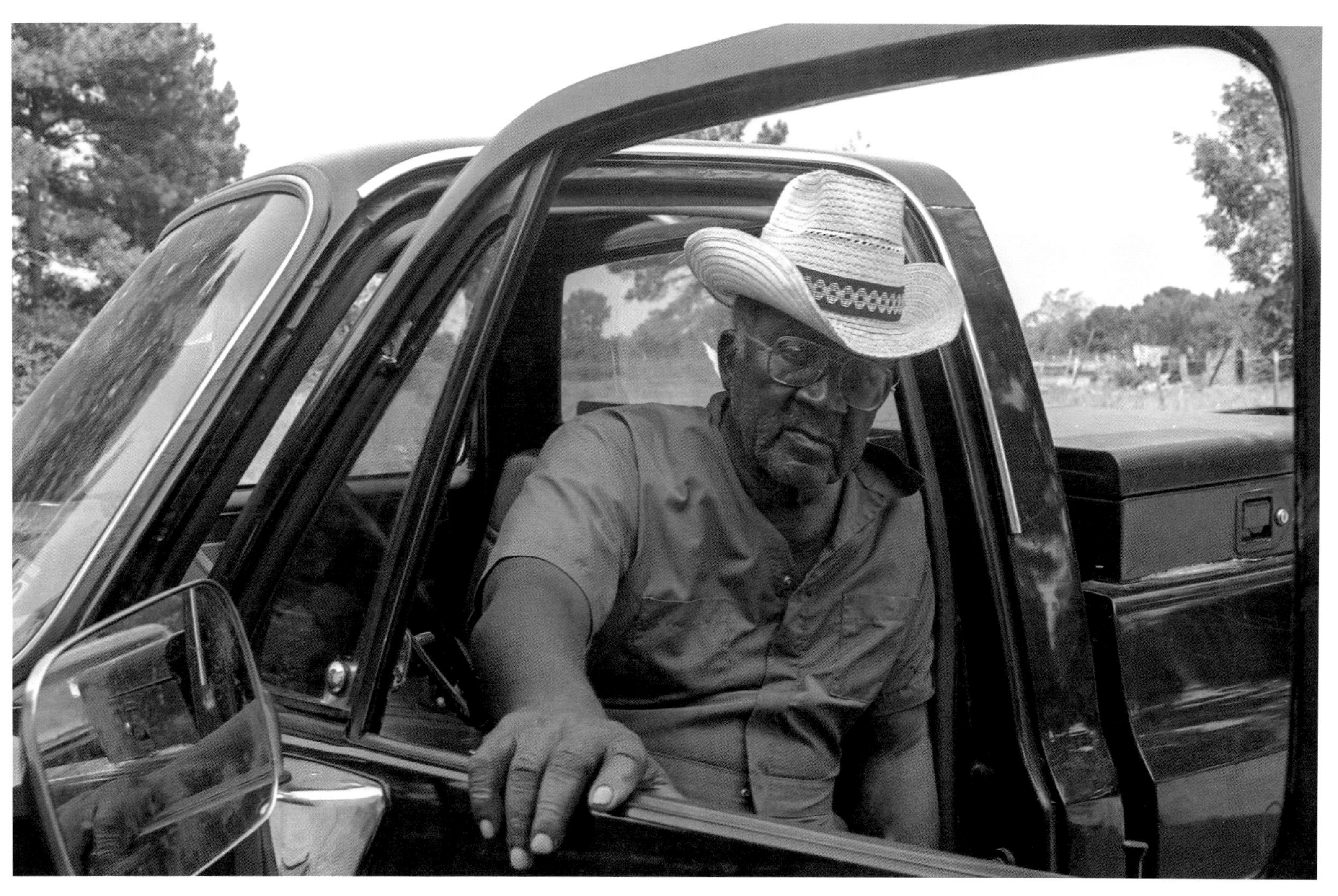

Monel Upshaw (1990)

My great-grandmother would stay with us. She would teach us a lot of stuff. She put a lot of emphasis on lying and stealing. She said, "Your name always goes ahead of you." What she meant was if you had a name of lying and stealing, somebody's going to tell, and when they want to know your reference they'd have it, "Can't believe nothing she says… you can't put nothing down." So I remembered that. Everywhere I worked people didn't have to wait to see me come in, and didn't have to be there when I left. They knew I wasn't going to take the house off . . . and I was going to be there on time.

—Leota Freeman Upshaw

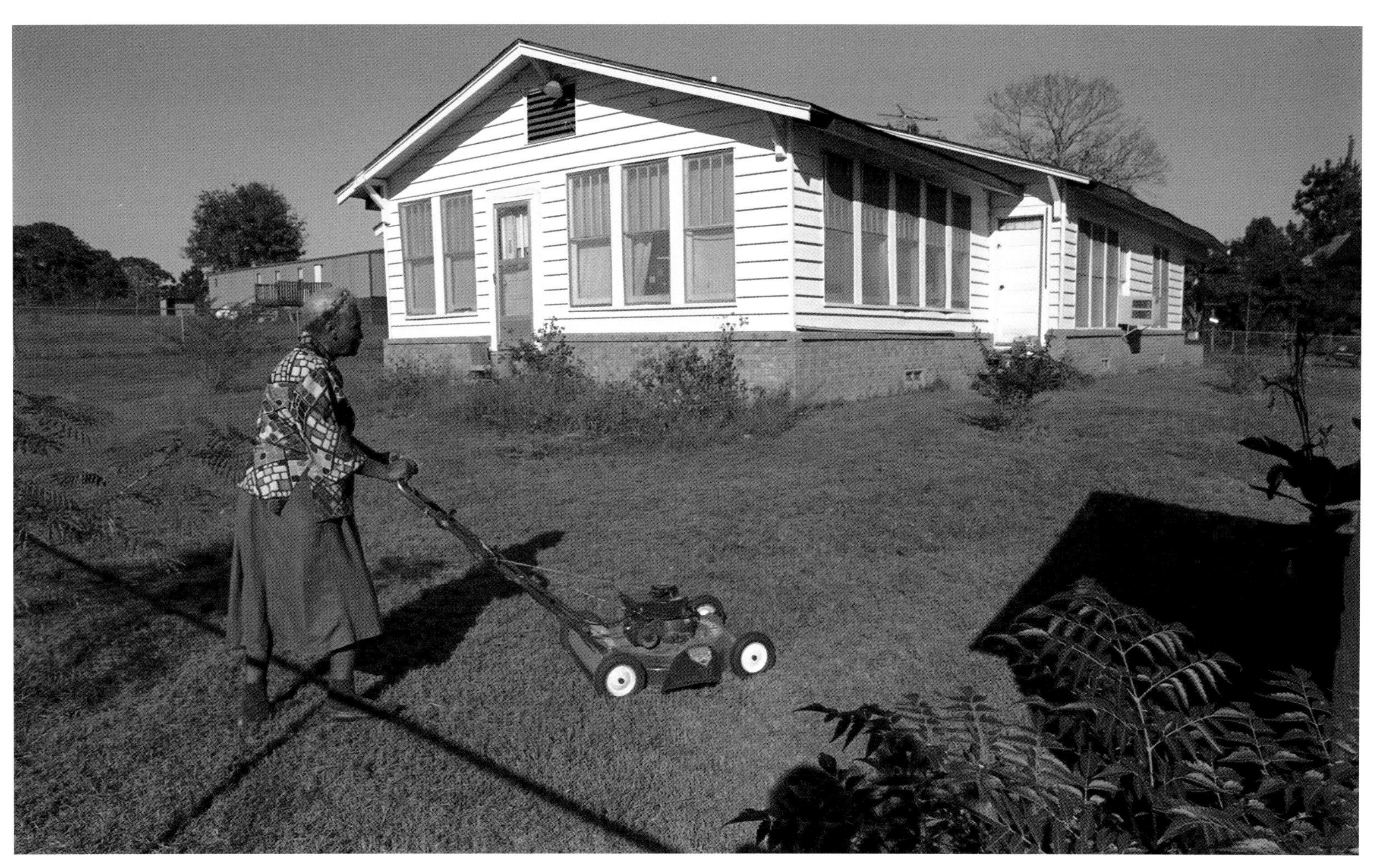

Leota Mowing (1990)

Growing up in rural deep East Texas certainly made for interesting times, though for many years I did not recognize much less appreciate them. I now realize just how good we had it. These days there are few places where children can roam freely without a passing worry about what evils might befall them.

There were no evils then, not in our corner of the universe. We were shrouded in ignorant bliss concerning what went on outside our haven. Far and away from us was the worry of drive-by shootings, drug busts, robberies and murders. Some of the terminology heard frequently today had not even been coined then. We were the innocent, naïve country folks that largely were the backbone of the South.

—Beatrice Upshaw

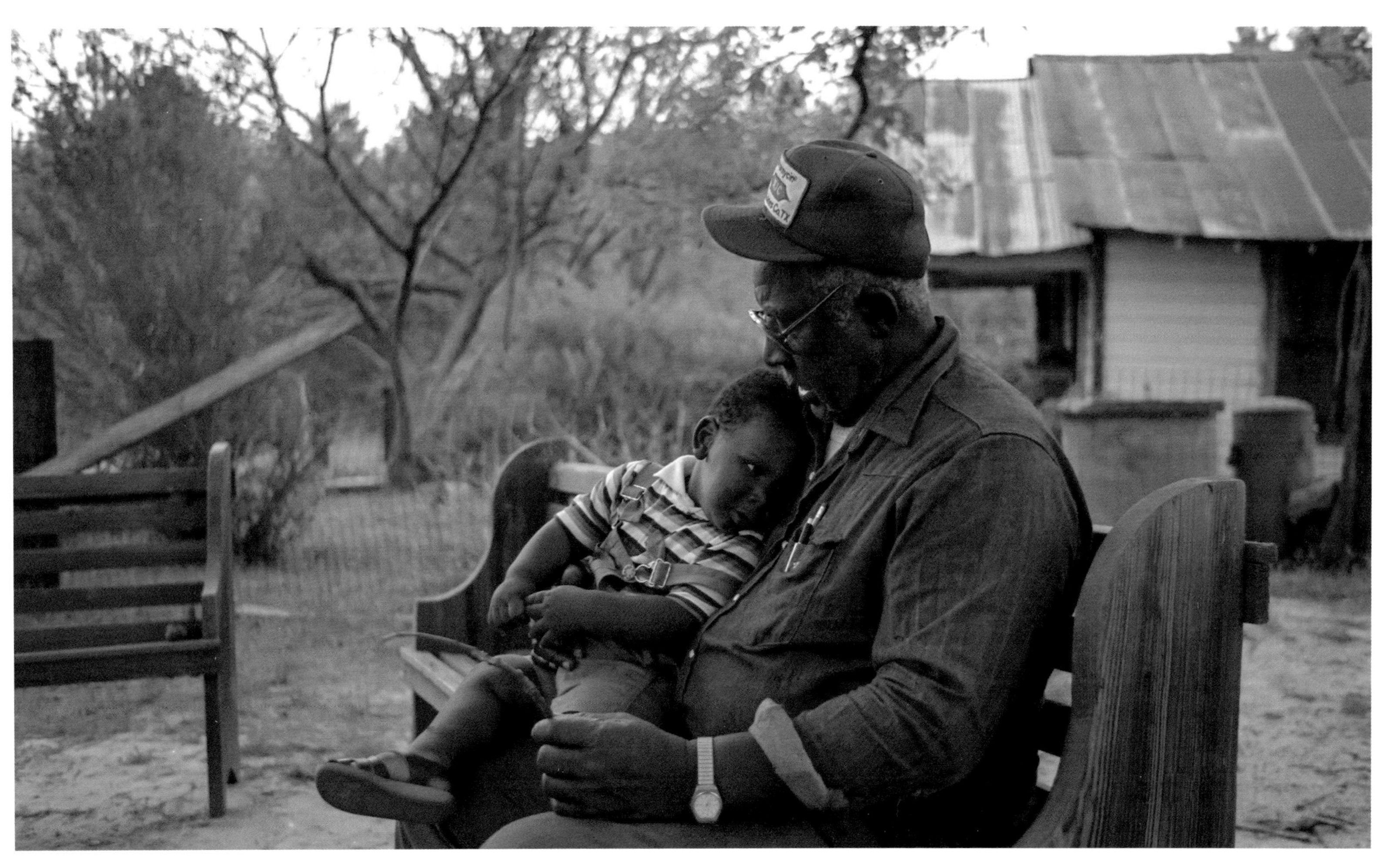

Grandfatherly Advice (1988)

She's probably sleeping . . . she might be praying, but she's probably sleeping. She would sleep on the sly.

—Beatrice Upshaw

Sometimes she would be asleep, but if you talk about her she'd wake up and start smiling. Even though she'd be asleep she'd still know what was going on all around her.

—Marceline Upshaw Harris

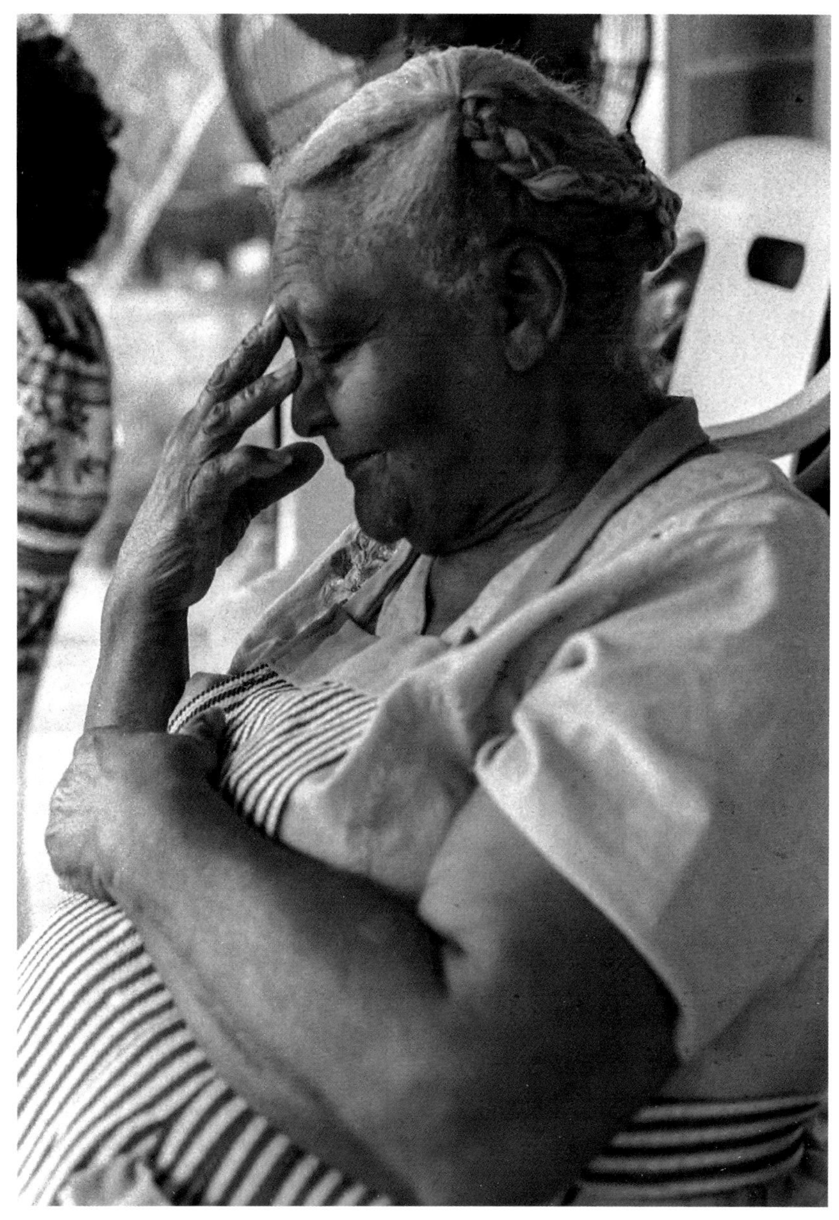

Leota Upshaw (1999)

As my sisters and I became young women and found prospective mates, my father issued the same stern imperative to each of our intended—his future sons-in-law. "Boy," he'd start, "don't you hit my daughter." With a stern look and a deadly serious demeanor Butch continued. "She's been raised once, and if you think she needs raising again, why you just bring her back home." He'd finish, never wavering his steely-eyed perusal of the young man's features. It was as if he were reading the intentions of his future in-law, literally reading the young man's mind while at the same time imparting a piece of his own.

—Maye Upshaw Ham

He made friends back when they used to play baseball . . . he and his brother . . . they had a baseball team. People in the community… they would go off and play ball with people in other communities. He made friends way back then that he kept until he died or they died. They came from everywhere around here . . . they went way back.

—Marceline Upshaw Harris

For the community he was a pillar. He was very comical. He had the reputation for making you laugh. He was very respected by white and black. That was how I knew how many friends my father had . . . when we had Homecoming at the church. My father had so many friends that I didn't know he had.

—Faye Upshaw Johnson

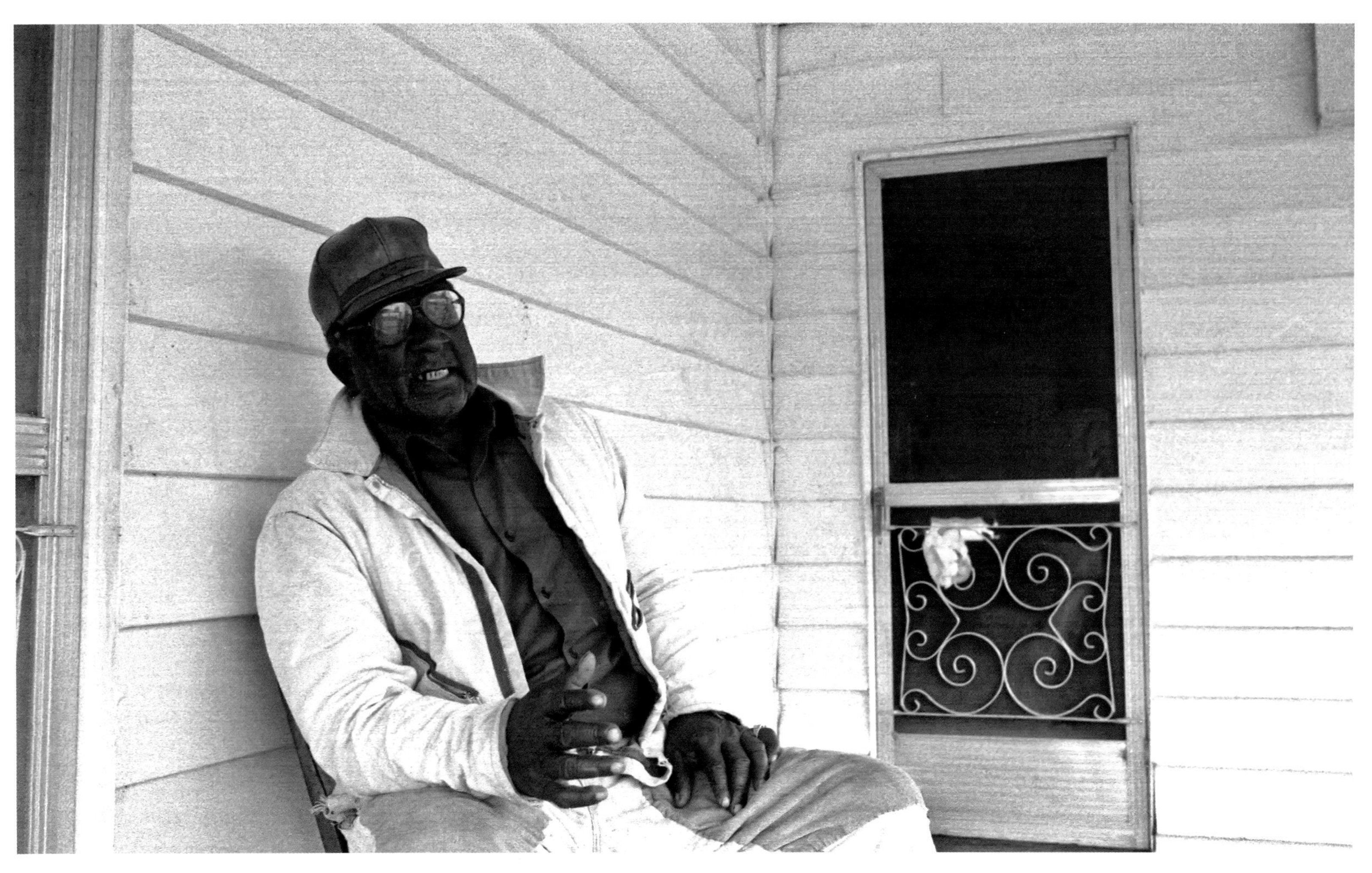

Monel Upshaw (1989)

No one had any trouble getting us to go to church. We'd get up in the morning . . . no one had to tell us to get up and get ready to go to church. My mother didn't just send us to church, she took us there. And my father was already there before we got there. We always thought that we survived because of our mother's prayers. She had a strong personal connection to God.

—Faye Upshaw Johnson

After breakfast we got ready for church. Our clothes were already clean and pressed; another Saturday chore since Mama did not allow us to wash or iron on the Sabbath. For some inexplicable reason, though, she didn't have a problem with our polishing our shoes on Sunday mornings.

Our black patent leather Sunday shoes were shiny due to their factory finish. Patent leather is always shiny but to really gloss them we used a biscuit leftover from breakfast. We took a cold biscuit, broke it into two pieces and briskly rubbed it into the leather, using one half for each shoe. The shortening in the biscuits made the shoes really glisten. The only problem was that if we walked to church the shortening also attracted the red sandy dust to our shoes. We solved this problem by always carrying a rag with us. Once we got inside the church we whipped out the rag and dusted our shoes off, refreshing that wonderful shine. And it was glorious.

—Beatrice Upshaw

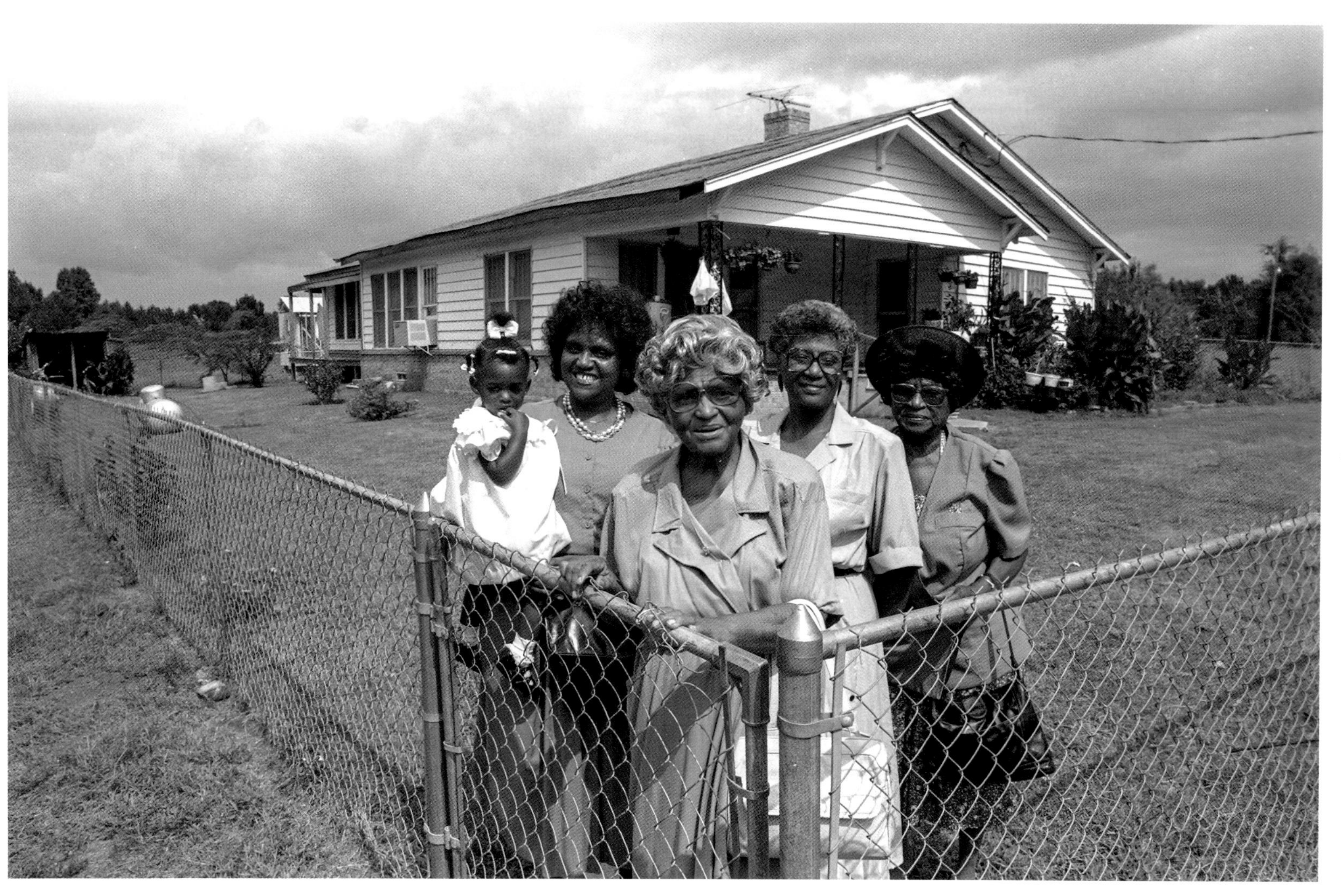

Going to Church (1991)

We had this little shotgun church out there . . . . I joined that church you know, not because I'd met God, but because they put me on something they called the "mourners bench." It was made out of slats that they got out of the log. So they sat all these sinner boys and sinner girls on this bench. You were on the mourners bench while they preached there in the front. Everybody was looking at us. The preacher was saying, "You're going to die and go to hell unless you join up." So that, plus the hard bench, caused us to join up that night in that little shotgun church. Our bottoms were worn out . . . . We all joined the same night… It's like it dropped down from heaven, the Holy Ghost. The preacher thought he was doing a good job, but we were getting off that hard bench. We didn't enjoy being preached to, because we didn't know we were that bad off. We were enjoying our lives the way they were. So we joined up. I guess that was a good move.

—Marion Upshaw

You know we'd look forward to going to church. We only had church one time a month. Somebody would go and get Reverend Dupree because he didn't drive. They'd go and get him in Alto on Saturday and he'd spend the night. He'd stay with somebody in the community.

—Marceline Upshaw Harris

Well we didn't have air conditioning, so the windows were up. And bugs would fly in and out. There were just fans. We had a string of lights outside… There was lots of activity going on outside. Monel was barbecuing, and we had a soda stand . . . bottles of soda in buckets of ice. That was really something. You could buy a barbecue sandwich and a bottle of soda water. There were as many people outside as in. And back then we had groups come from Louisiana. That was a big thing back then. They came from Shreveport . . . singing groups, quartets. Those were good days. We had better all night singing back then than we do now. It was definitely a social event.

—Marilyn Upshaw and Beatrice Upshaw

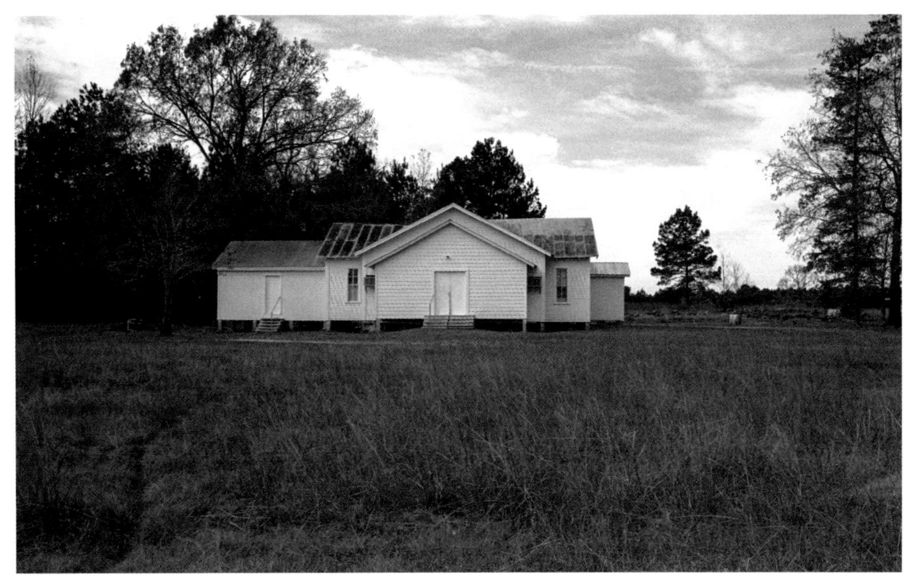

County Line Missionary Baptist Church (1990)

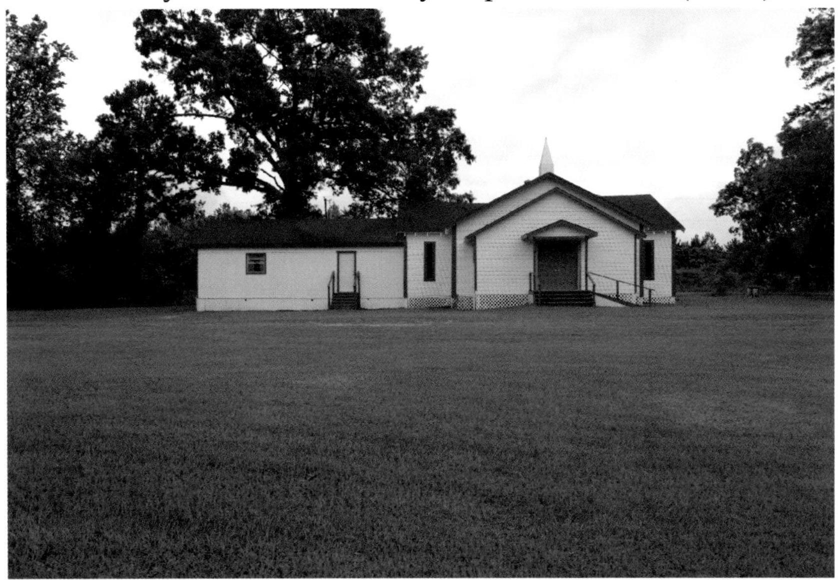

County Line Baptist Church (2009)

The preachers walked in here. They walked in and they walked out. The preacher I remember was T. D. Dupree. As far as earthly angels were concerned in a man, he was one of them. He would walk in here, or catch a ride on a Saturday evening, spend the night with one of the brothers or sisters, get up the next day and preach. I would see him stand over the pulpit while the brothers were taking up the offering, and ask, "What you got there brother?" "Thirty-five cents." "Well give me a dime." I never forgot that man. He walked over here from Rusk, and farmed during the week. I never saw him take more than a dime. Sometimes he would take nothing, or a chicken.

—Dolvin Upshaw

If you've been in the field all week you'd be glad to go to church. Where else were you going to go? There wasn't anything else going on. Didn't have TV's to come back home and watch the ball game. [Going to church] was something to really look forward to. We made friends . . . pure type friends. You see some remnants of this happening now.

—Marceline Upshaw Harris and Claude Upshaw

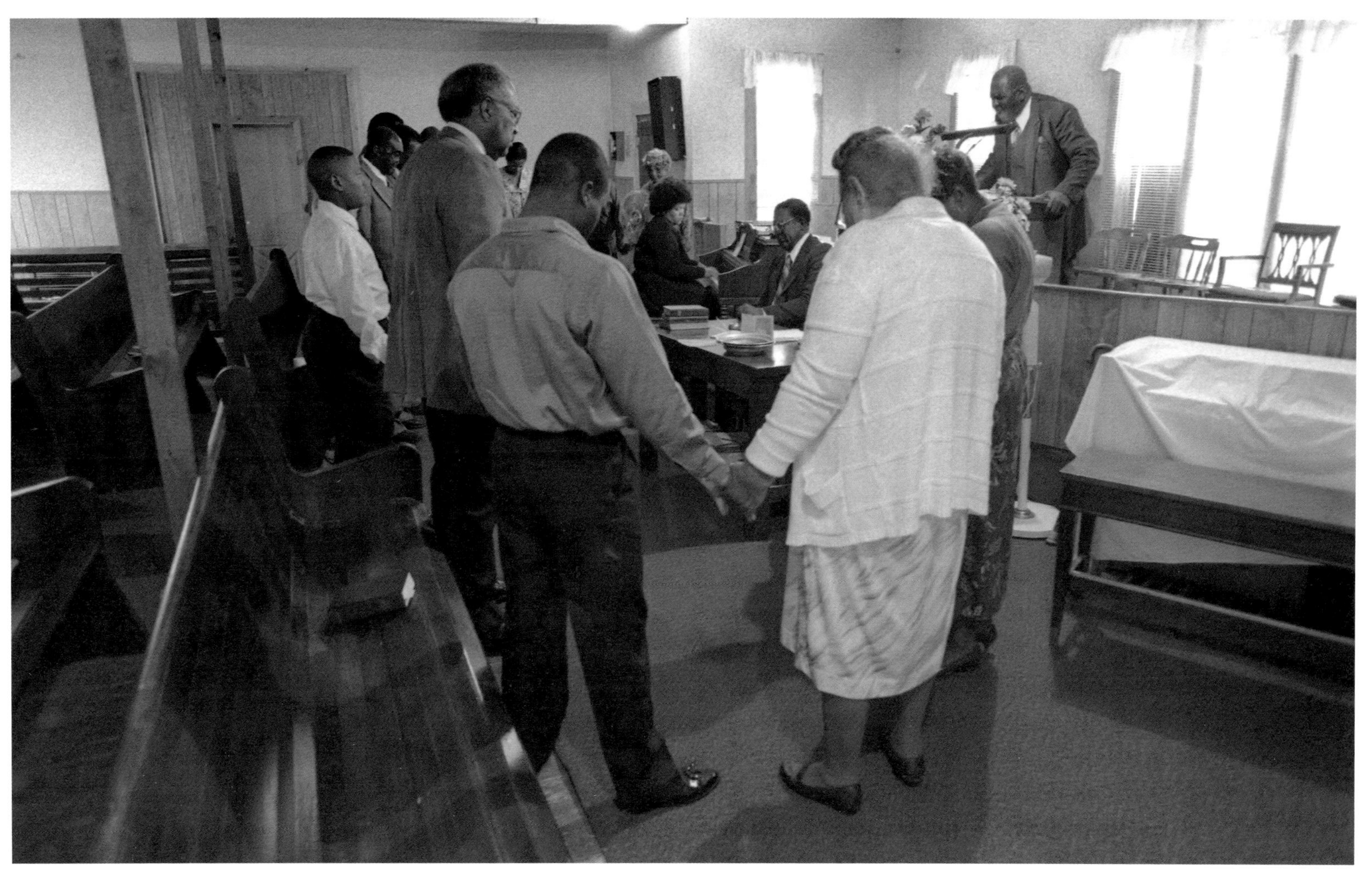

Altar Prayer (1990)

The days passed swiftly from the time school dismissed in May until the first big event of the summer. The annual All-Night Singing was held at our church on the second Saturday night in July. For the occasion neighbors and local church members were invited to spend the evening with our church family, singing and praising the Lord. Sometimes the singing went on until well after midnight . . . thus the title of the event. This was a time of great excitement for us kids. There was not a lot else to do in the community. School was out and we could only spend so much time hanging out at the store. Going to church was a great social occasion and relief from the boredom of summer in the country.

The All-Night Singing was the only event of the year which required the church campus to be rigged with extra lighting. The bare one-hundred-watt bulbs were strung about the churchyard, hanging perilously from electrical lines, which had been attached to trees and poles. The bulbs provided ample light for the outside area. They also attracted lots of bugs. There were bugs inside the church as well since the open windows were uncovered.

After a group sang their selections they often meandered outside for fresh air and a cold drink. It got mighty hot in that old un-airconditioned church and the only way to cool off was to step outside. Singers, individual and groups, strolled outside the sanctuary. They mopped their brows with already sopping handkerchiefs in an attempt to find a modicum of relief from the stifling heat.

We did not mind the heat. The excitement of the service more than offset the discomfort we endured. These were high times in County Line and, next to the store, the church was our favorite place to be. Admittedly, our religious experience was not what it should have been because we were so very intent upon having a good time. Even so, we did have fun at church, albeit surreptitiously.

—Beatrice Upshaw

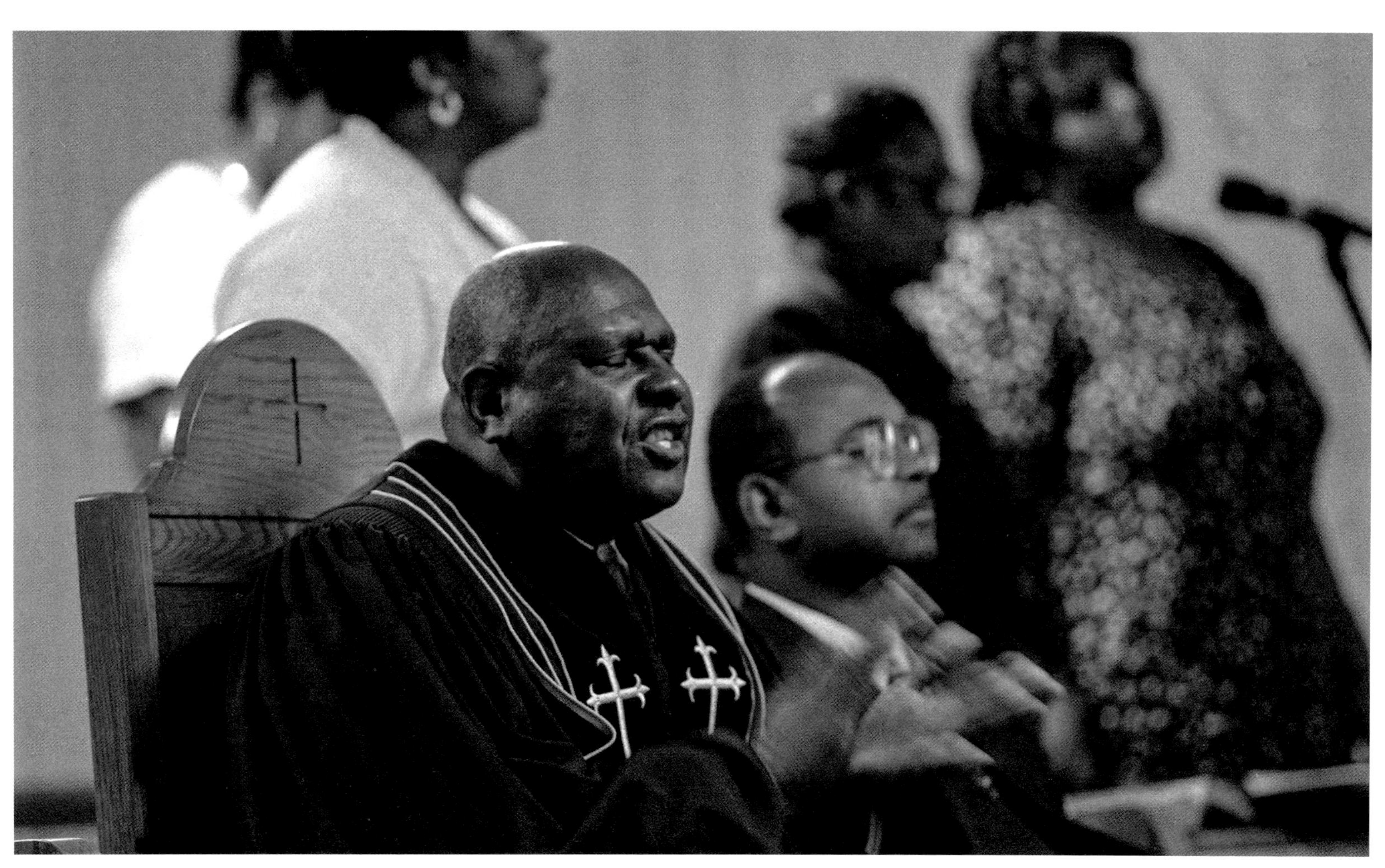

Preacher Wickware Clapping (1997)

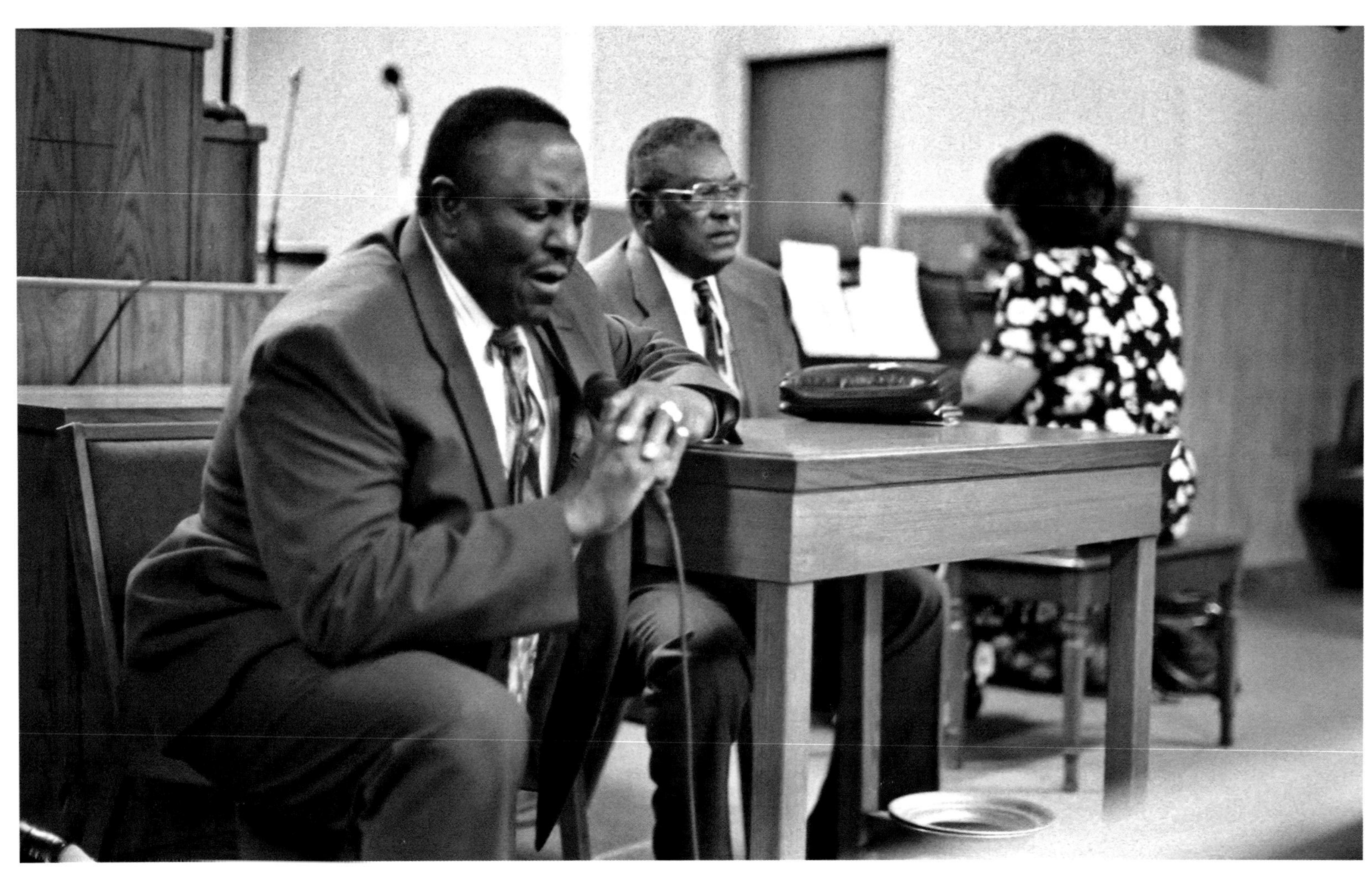

Hady Praying (1998)

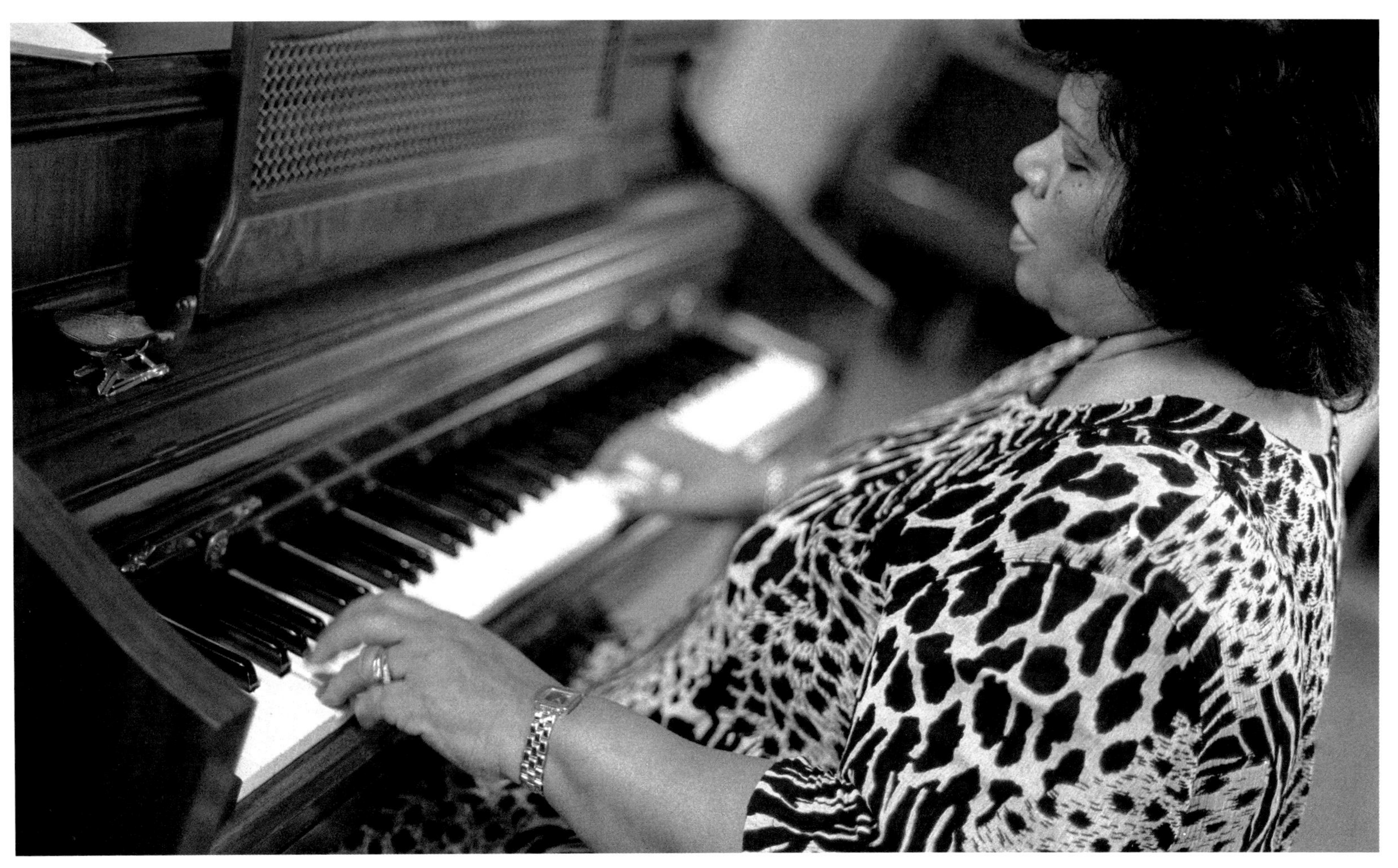

Maye (1997)

What we now call the Homecoming we used to call the August Meet. Everybody who could come back home would come back, just like now. That's been going on for years, as far back as I can remember. We would call it the August Meet. We didn't have the cafeterias that we have now, so what they would do, they would put benches outside. And all the families would put their food there. And some of them would have it on the backs of their cars. The guests could go from one to the other. "Give me some of your corn, give me some of your pie." The church grounds would be full. The church couldn't hold them all. They'd all come back from California or wherever else they migrated to. All the neighboring churches would come and participate. And those were some real days.

—DeMorris Skinner Young

The Second Sunday of each month was a big thing. It was a time when [the women] would cook their best meals, bring the preachers home to feed them, and since we didn't have telephones you got to learn more about what had happened in other places and to other people. It just didn't happen often enough, once a year, but it was something I know my mother looked forward to. And I can remember that she would cook, and the other women in the community, would cook their best the Sunday of Homecoming. They put [their food] in a big box and took it up to the church and put it out under the tree.

—Faye Upshaw Johnson

People would come from miles around. There'd be so many people out there you couldn't hardly walk in the yard. Everybody didn't go in the church because everybody couldn't get in there. You'd walk around and communicate with each other. They'd be parked all the way up the road to Aunt Louisiana's and all the way back. But then that was the way it was at all the country churches back then.

—Marceline Upshaw Harris

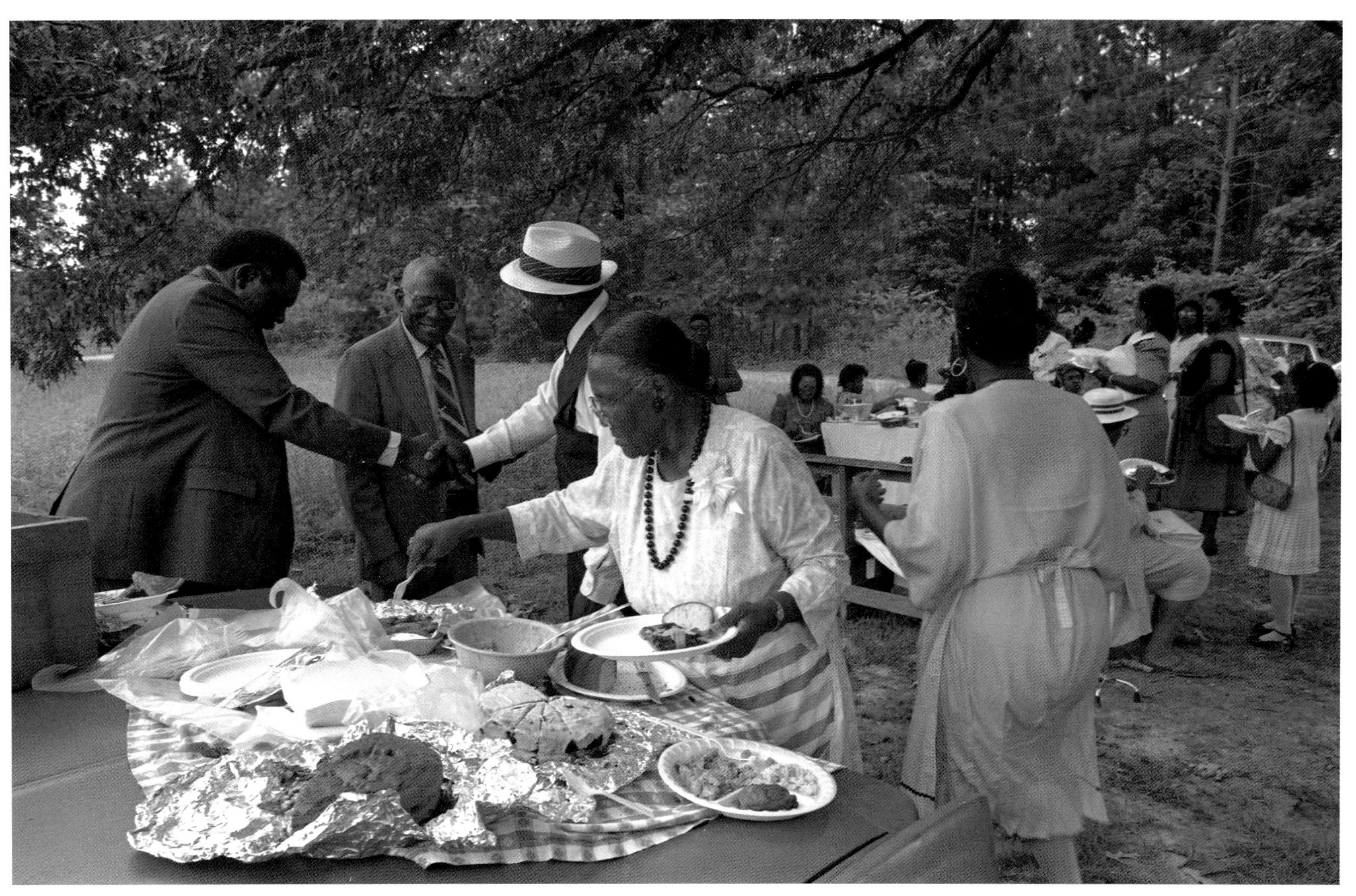

Dinner on the Grounds (1991)

We used to all gather around the piano when we came in for Homecoming. We didn't call it that then . . . we called it the "August Meet" . . . that's what we used to call it long time ago. We'd all gather around the piano when we came into town. We'd have a singing jubilee.

—DeMorris Skinner Young

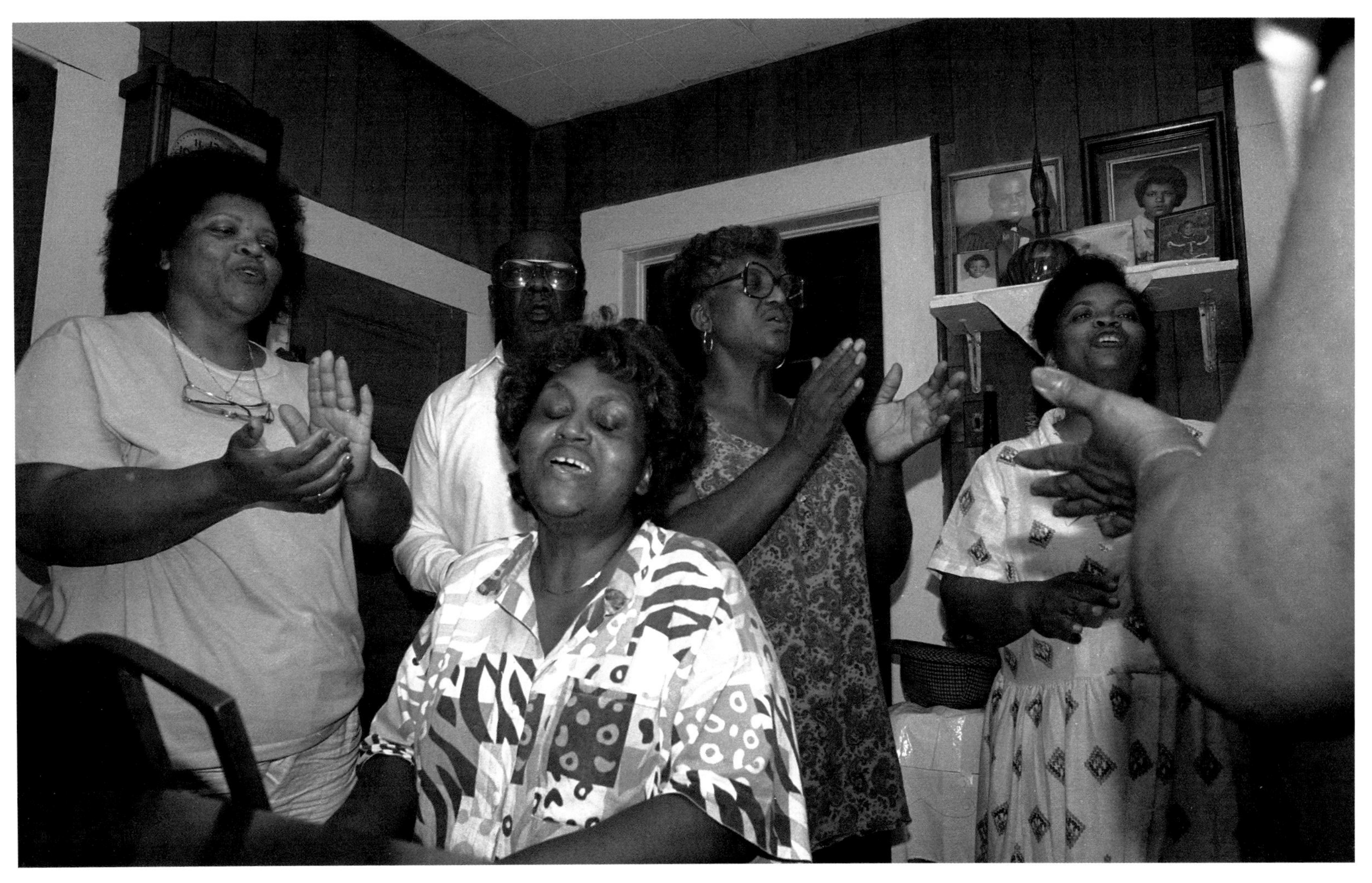

Family Sing (1991)

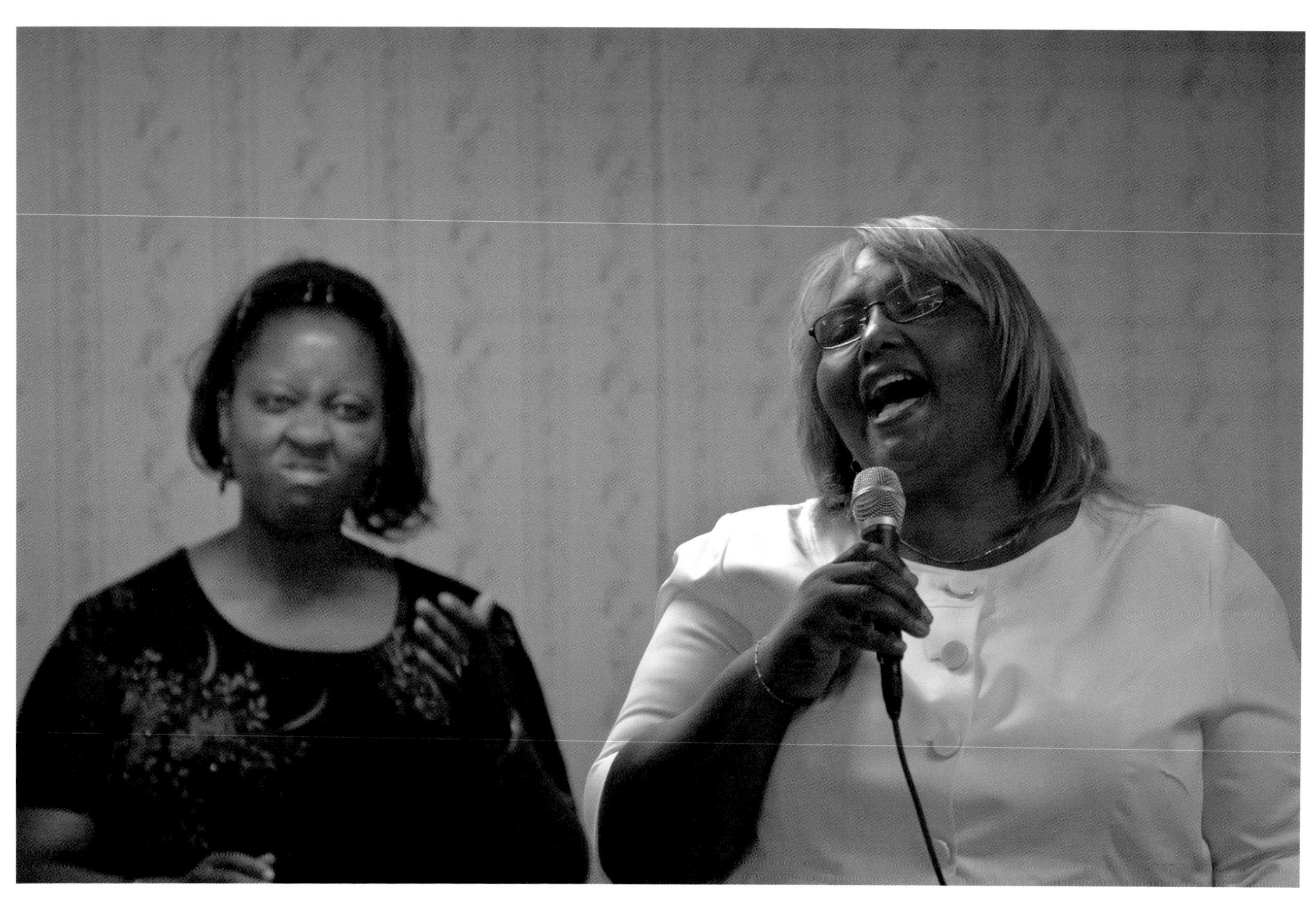

Terí and Marilyn (2009)

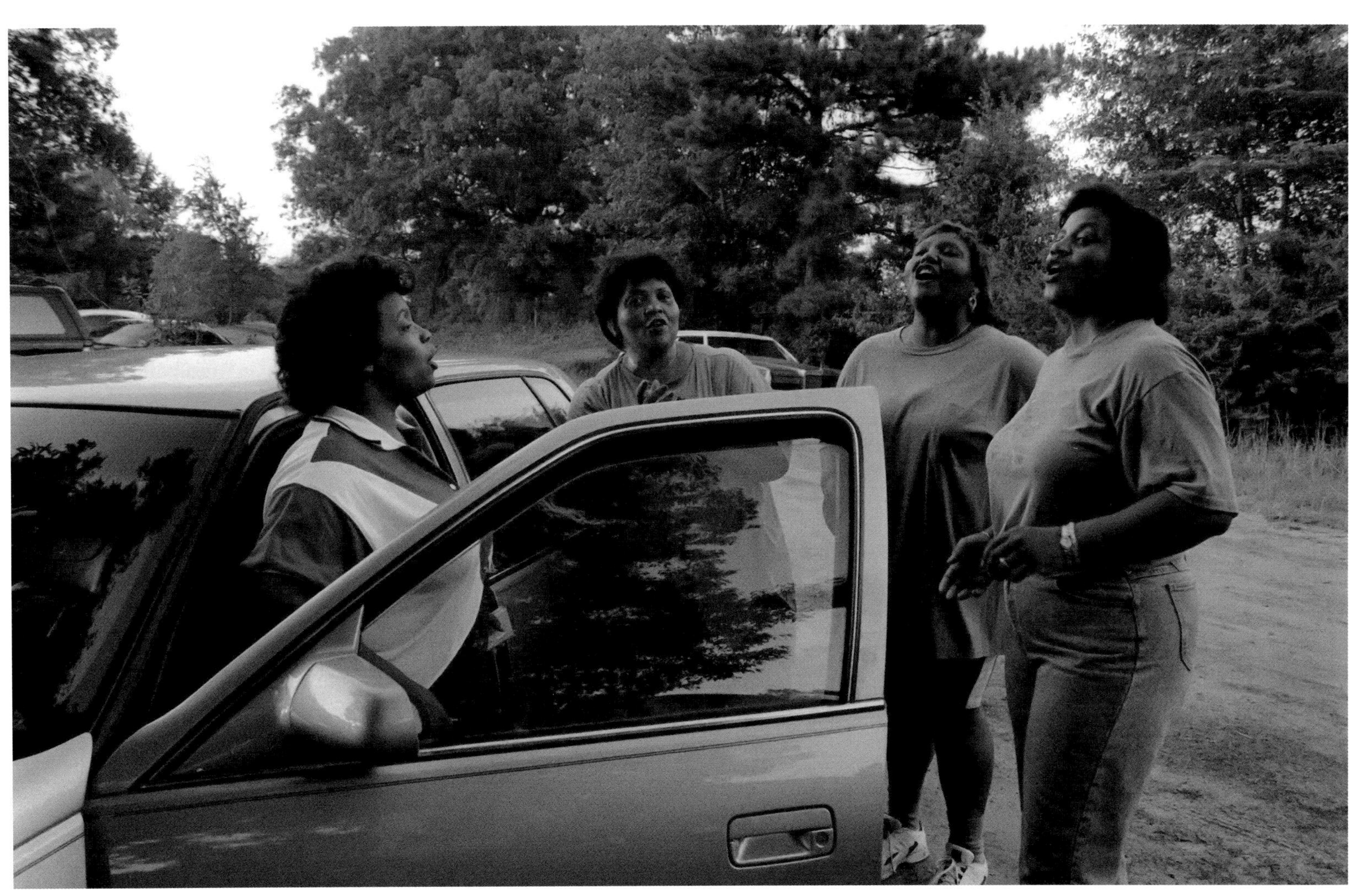

Faye, Maye, Marilyn, and Beatrice (1997)

Homecoming evolved into the gathering under the tree on Saturday. It just got larger and larger and larger. As far back as I can remember people did come home especially on that second Sunday in August. That was our biggest gathering by far. When they gave it the label family reunion, then it started to mushroom.

—Faye Upshaw Johnson

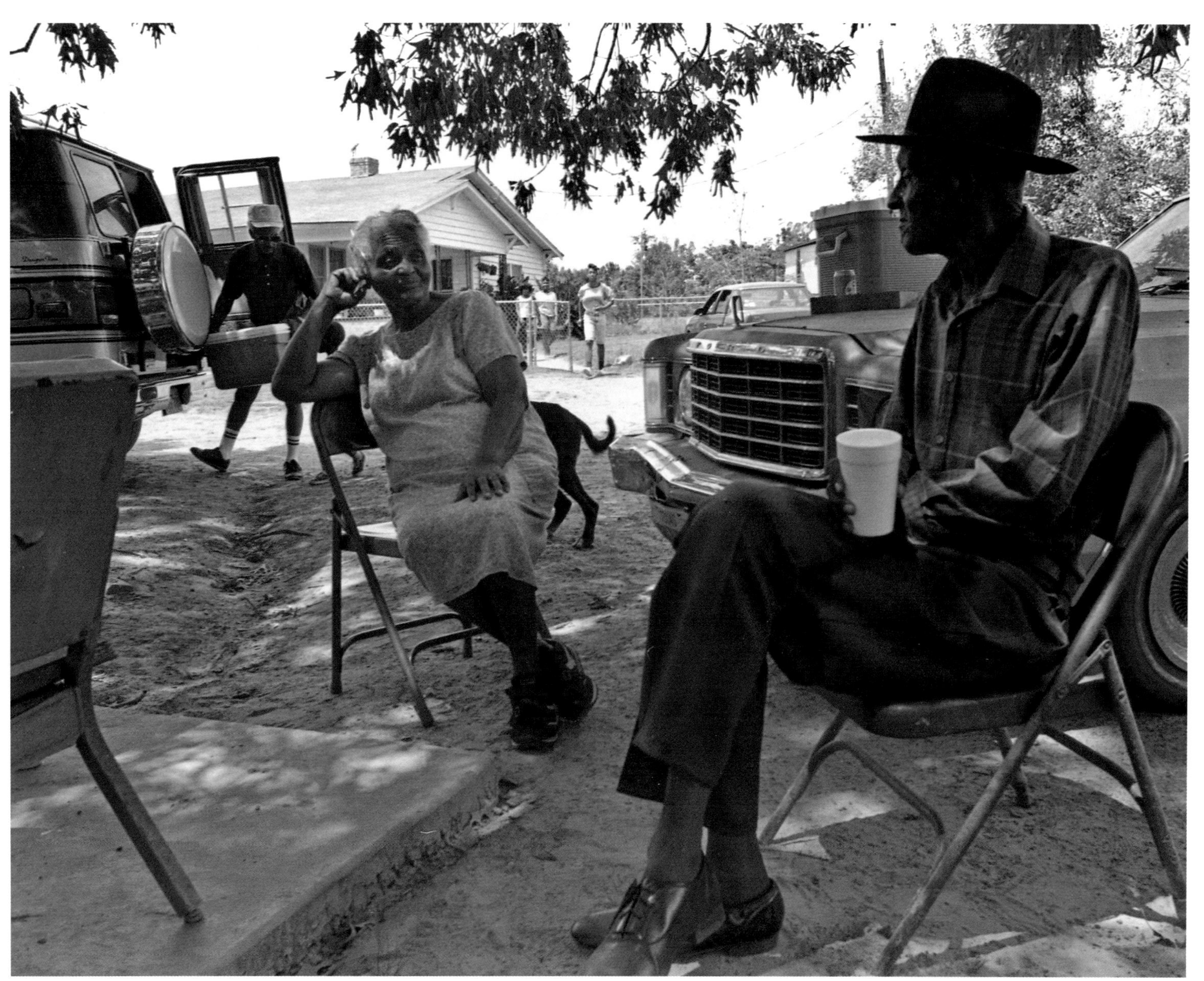

Leota and a Visitor during Homecoming (1991)

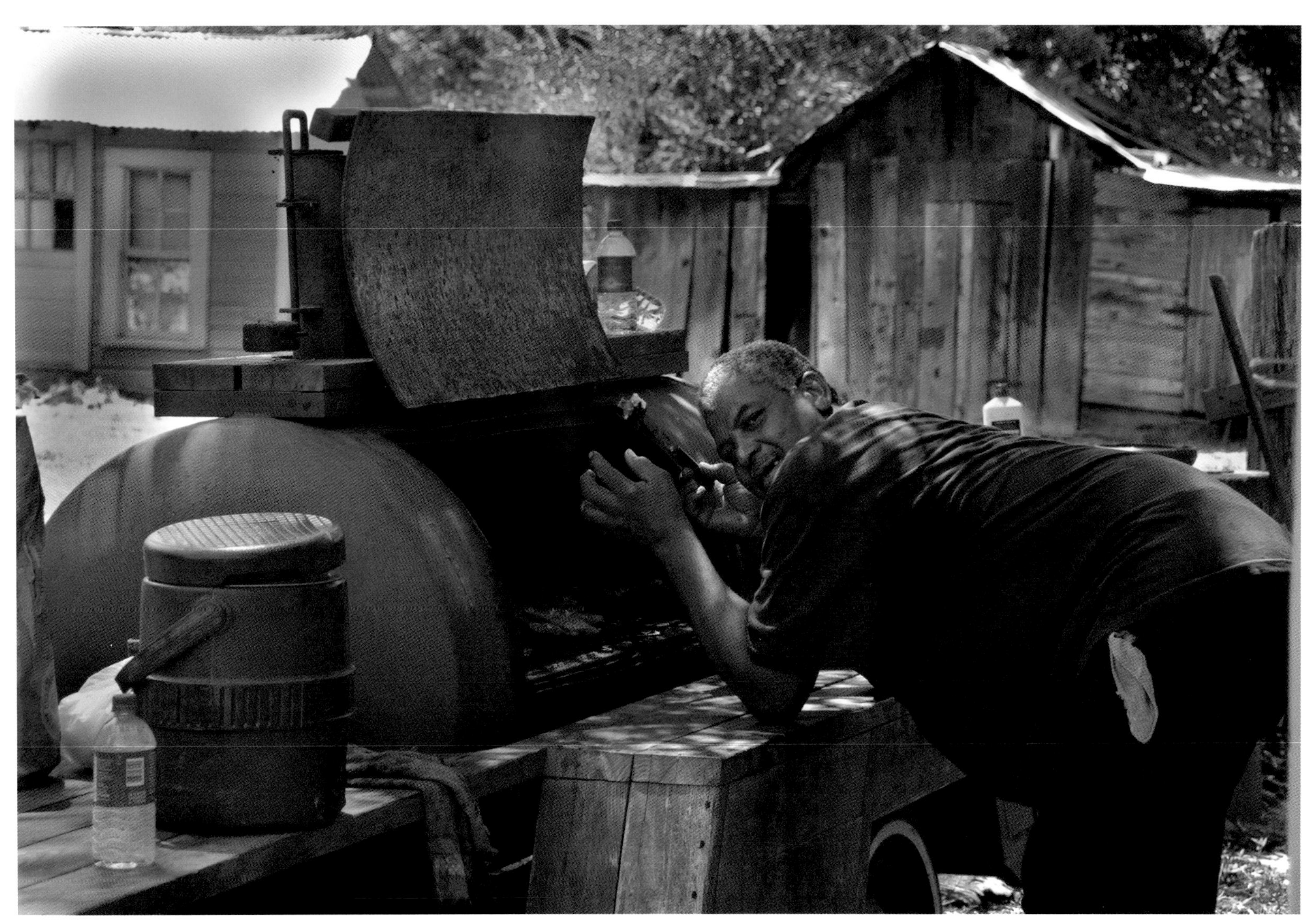

Pete at His Pit (2006)

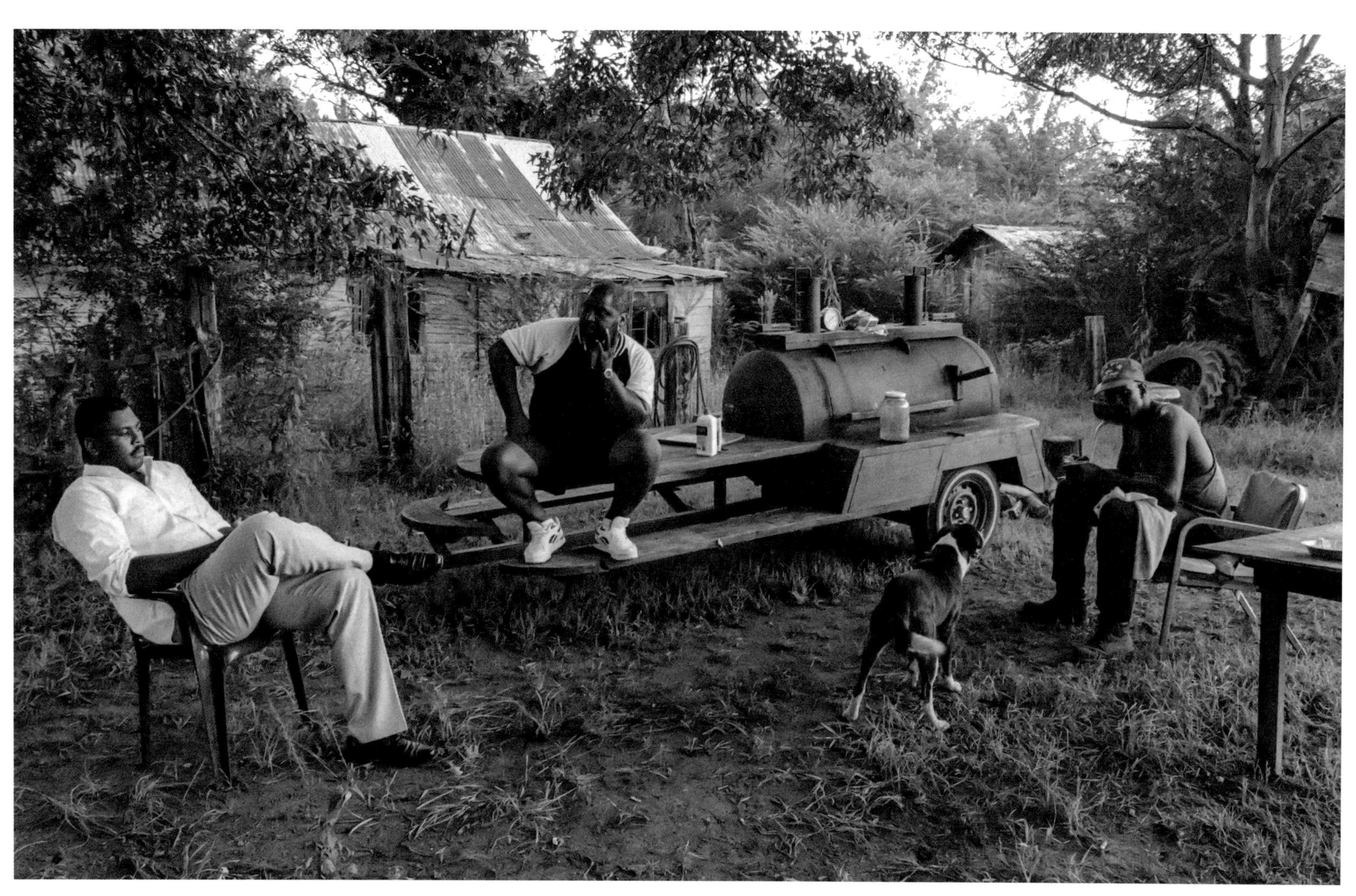

Al, Gus, and Soldum (1997)

My earliest memories also include when the new child was born. My mother would not tell us how a child was born. She always told us that she went out in the back of the woods and found them under a rock. Or a log. We really thought that that's how the babies got here. We were always happy to have a new addition to the family. My mother and father never ever talked about having too many kids, or the struggle of having all those children. Whenever a new child was born we just put a little more food on the table, and there was enough love to go around. So we always loved the little babies, and in the next year and a half or two there would be another one.

—Faye Upshaw Johnson

Mama used to tell about the time she left this cake at the house, and when she came back all the icing was gone. Naturally nobody knew a thing about it, so she said, "That's okay, I put some poison in it, but I've got this medicine here, and if they don't take it, they're going to die." It was blueing water that she had, so everybody lined up to take the medicine. All of us were guilty.

—Maye Upshaw Ham

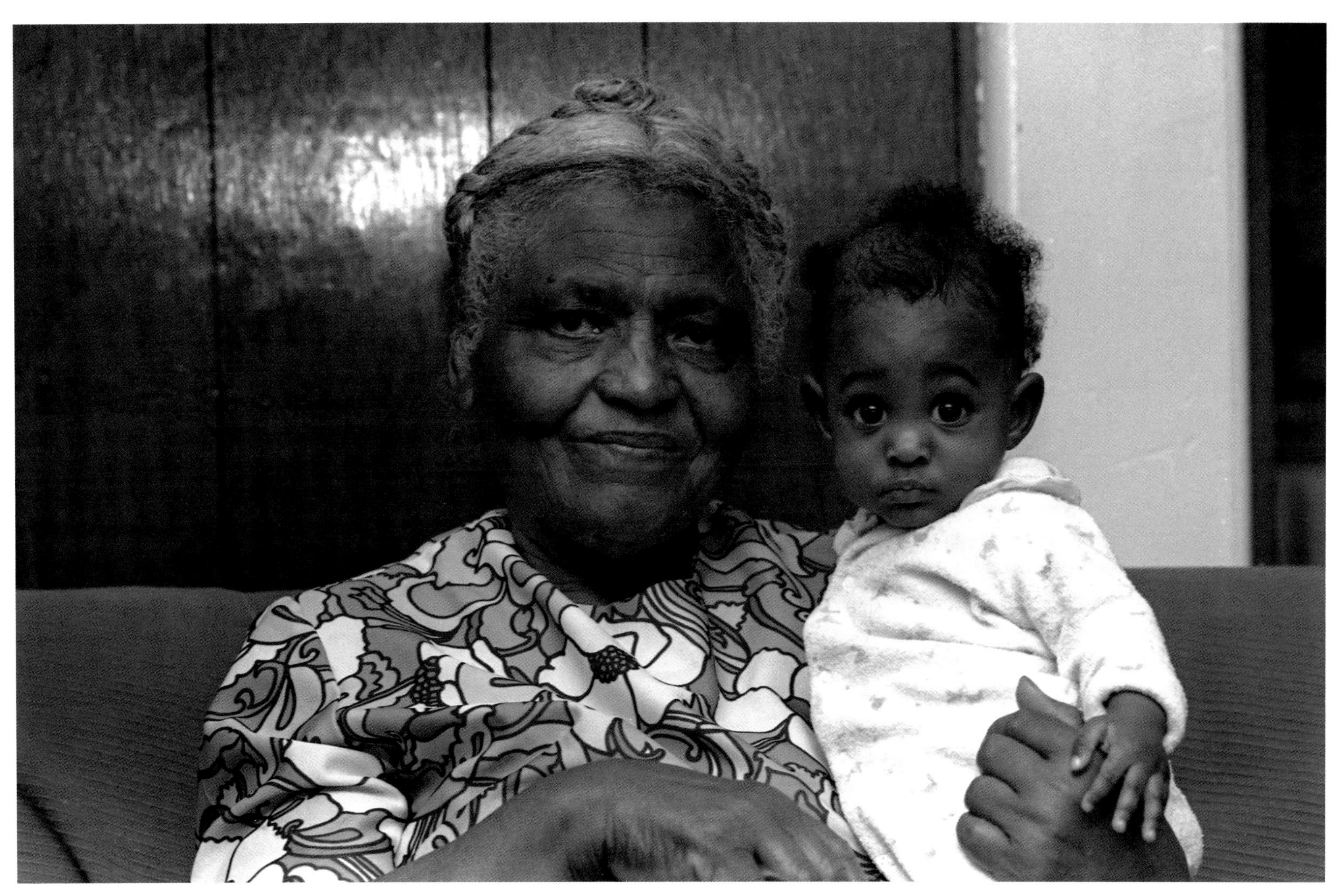

Leota and Sherrell (1990)

I lived with everybody. I could go to any house [in the community] as if it were my own . . . it was a real community. I don't think any of my friends or classmates knew anything about that. I was about six or seven when I realized this and understood that I could not talk about these things outside my family. [None of my friends] could relate to that. With my friends, they had a mother and a father and some aunts and uncles who lived a long way off that they only saw on Christmas or holidays. And they stayed in this one place, this one house . . .

For me, I did not have just one home. I could go to California and stay with Aunt Faye and feel as at home as I would back in Texas. The family dynamic was different. I did not have just one mother, I had six or eight, counting my aunts and grandmother. On family days at school my aunts and uncles would come along with my mother. We were like gypsies in that sense. Yeah, it was different.

—Elia Upshaw Ali

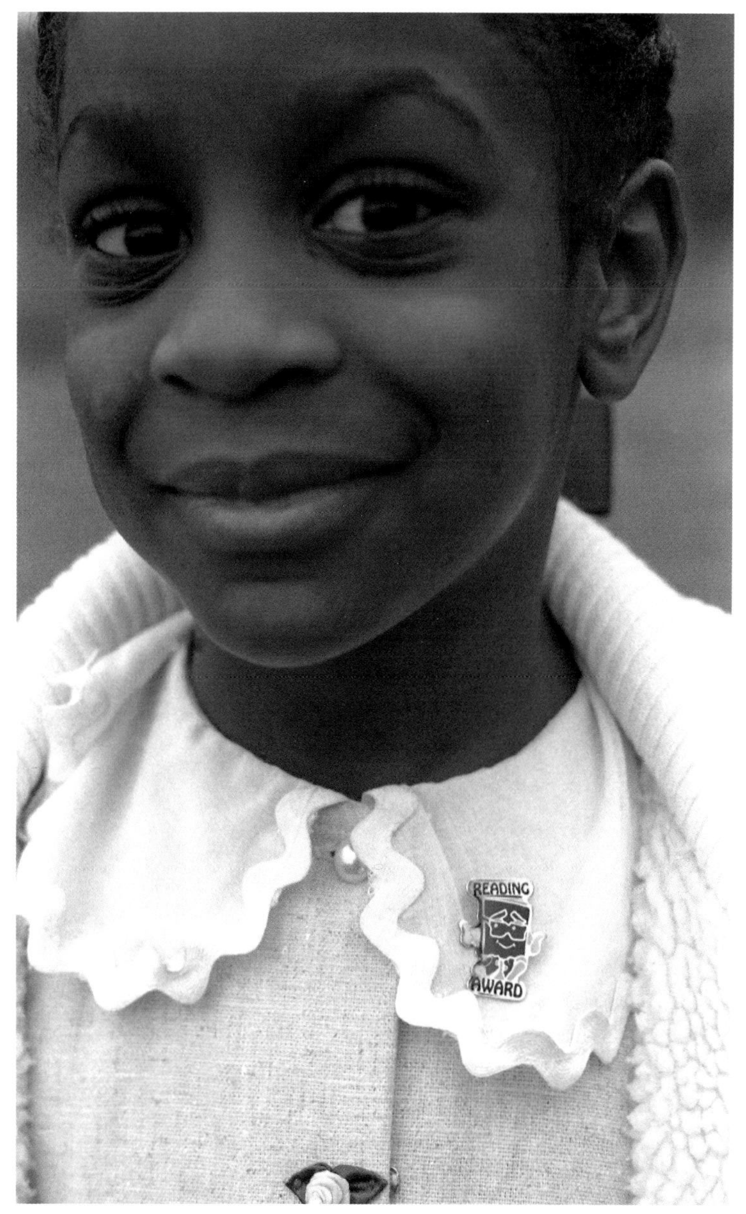

Elia — 100 Books (1997)

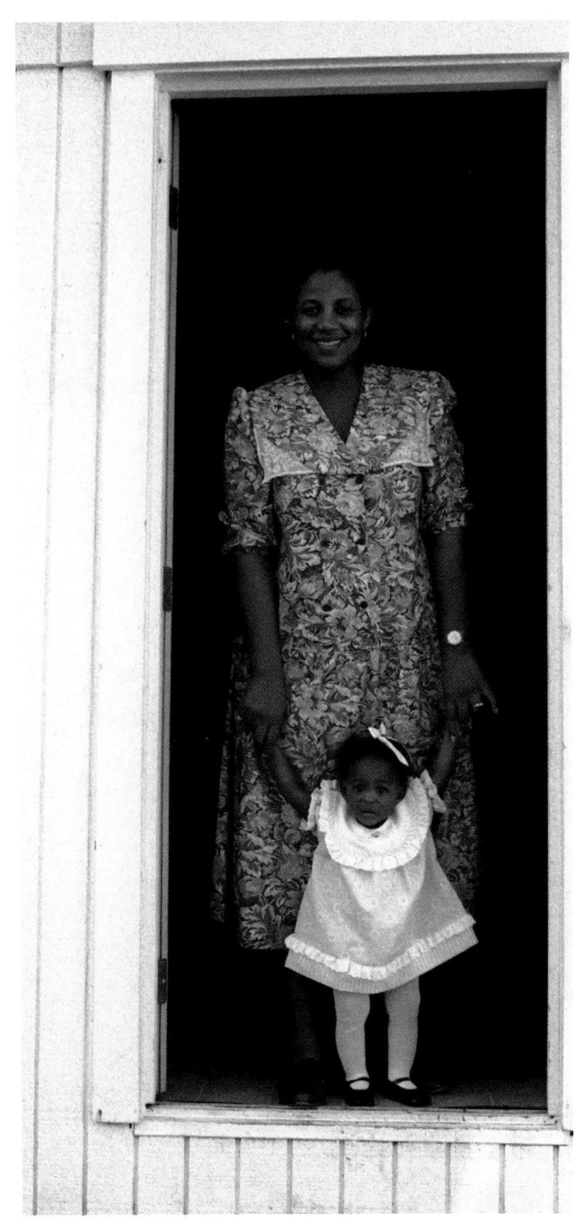

Beatrice and Sherrell (1990)

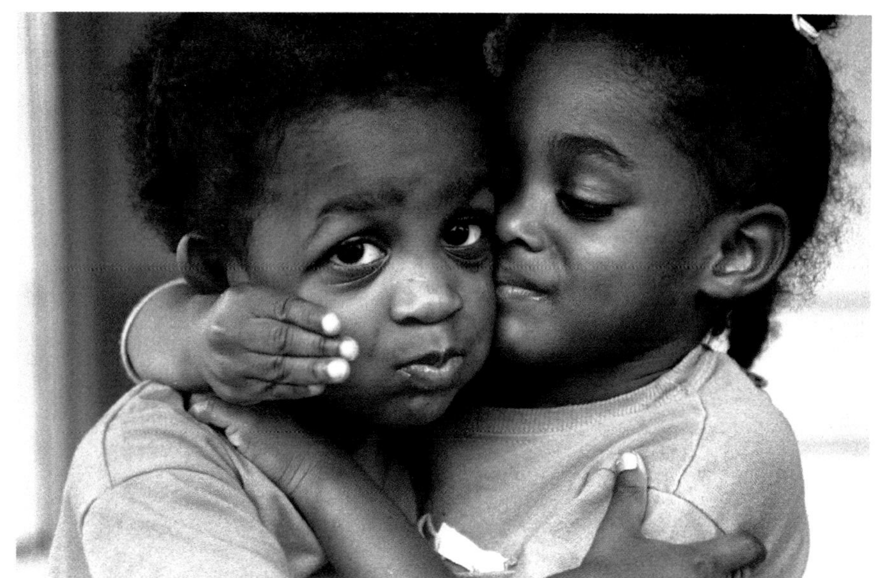

Elia and Sherrell (1993)

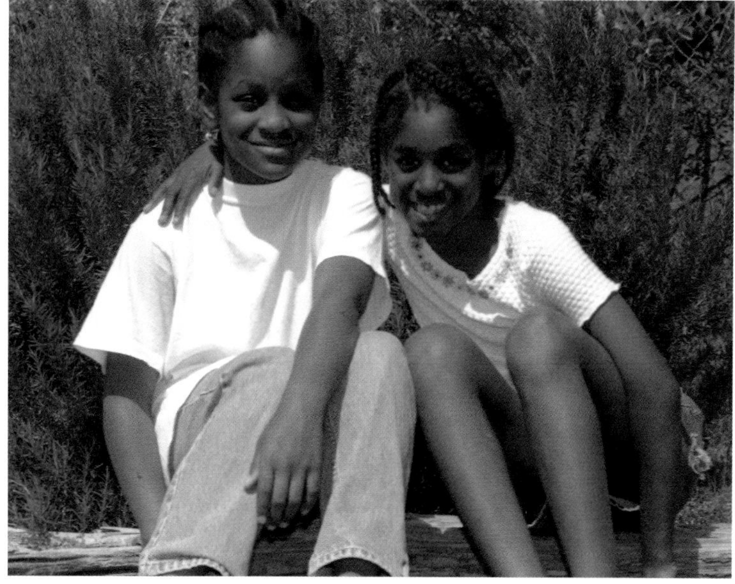

Elia and Sherrell (2000)

My mother was a songster and I can remember that she would always sing. When I would go to school there was a store nearby and my Uncle Claude and Cousin Gillis and some of the other men they would be sitting around the heater and I would stop by the store and they would say, "DeMorris sing us a song." And I would sing for them proudly and they would give me a little change. I developed a love for music because of my mother. At recess time instead of me going out to play I would sit at the piano and pick out tunes and I taught myself how to play but I played by sound.

—DeMorris Skinner Young

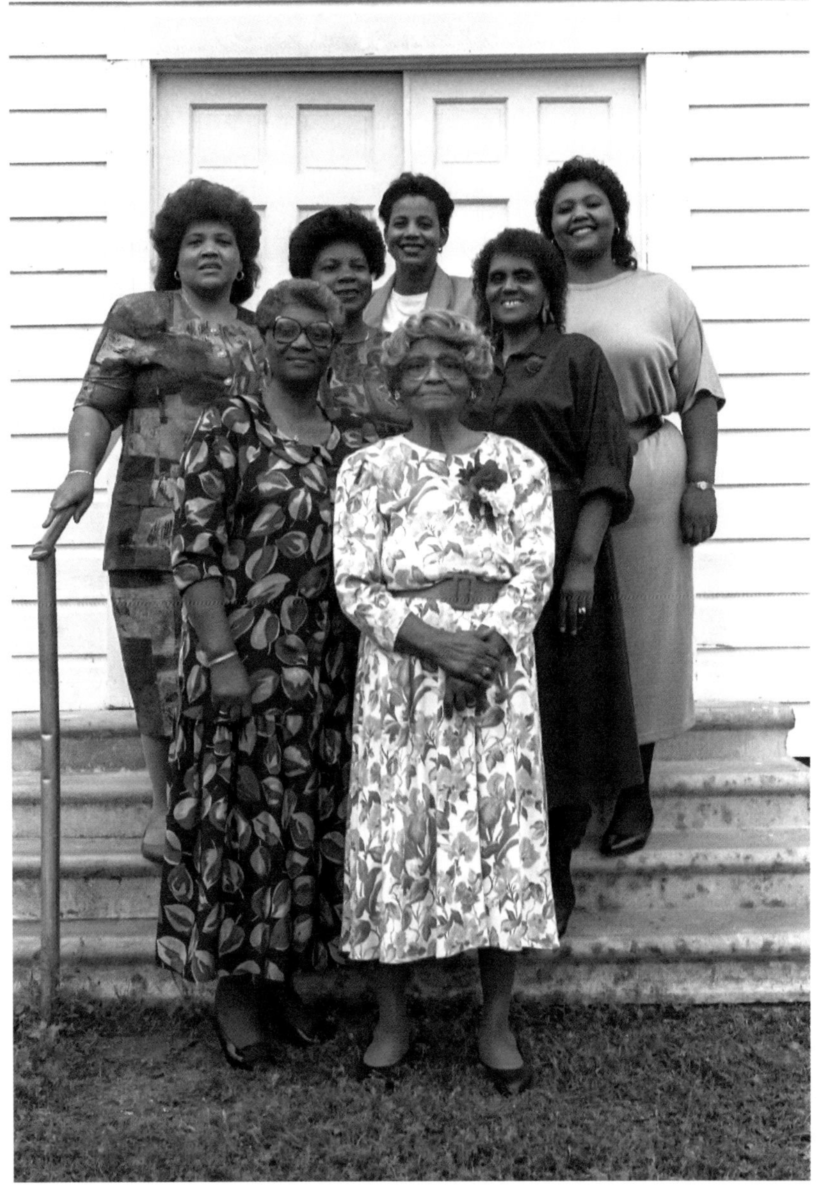

Mother's Day (1993)

My mother and father were very big on education. It was very important that we got through school. And we did. I remember hard times, and I remember struggles, but I never remember my mother or father complaining. It was just something that we got through. And if I had to change it, I don't know if there's anything that I would change. Not anything.

—Faye Upshaw Johnson

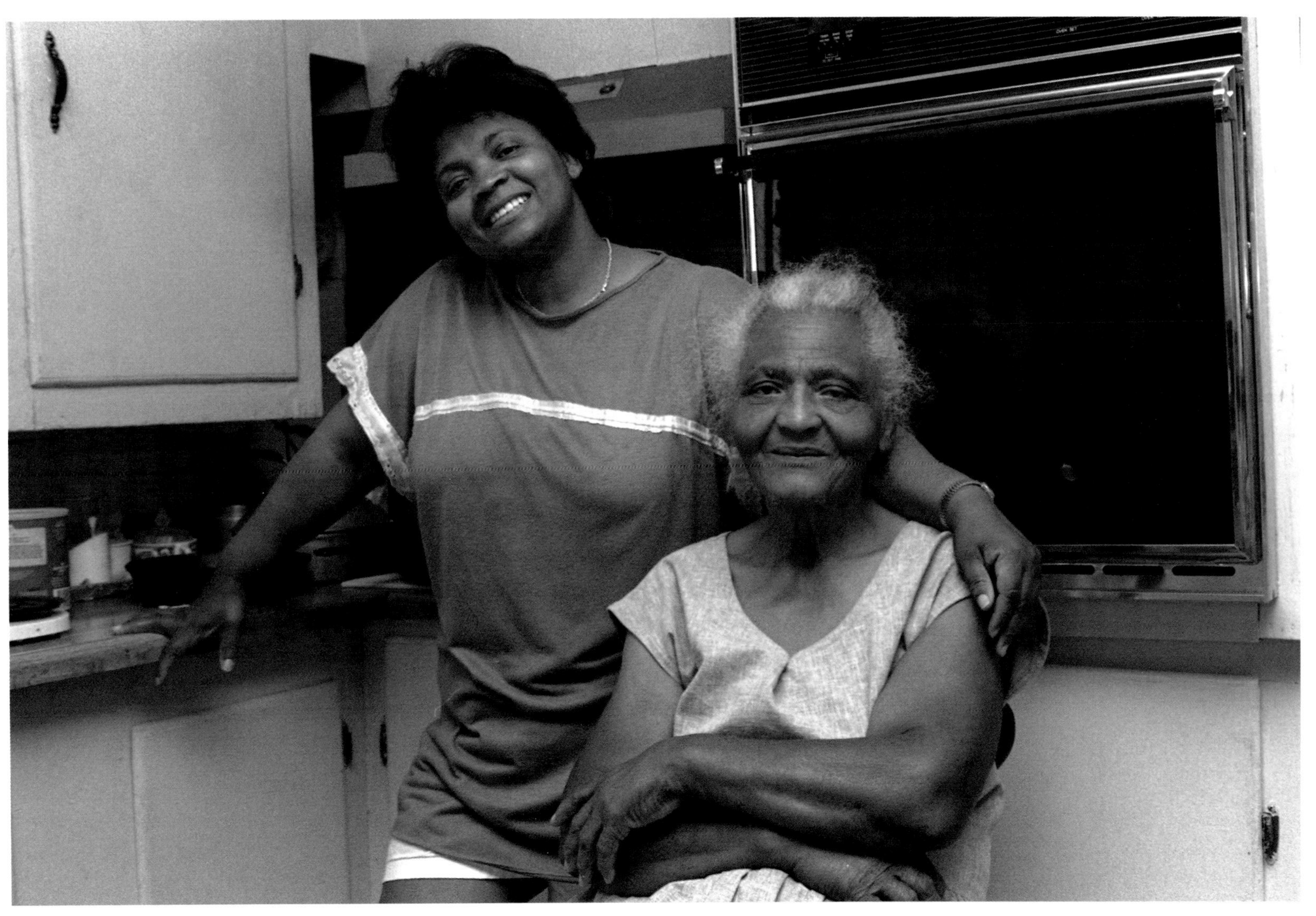

Faye and Leota (1994)

They (our parents) taught us that education was the way out of the fields. My daddy had a saying, "Fix your head and your head can fix your feet." By that he meant, get an education and you can buy all the fine clothes you want. Another thing he said was, "Boy, don't treat your head like it's just a block between your ears."

—Dolvin Upshaw

Going to school was a day for celebrating. If you went to school it was better than going to the field. So going to school was a grand privilege. I welcomed the coming of the rain, because that meant you were going to school the next morning.

—Marion Upshaw

We had Interscholastic League activities which gave us a chance to go to Prairie View about three different times. We won gold medals. Mine was in spelling. We had spelling, arithmetic, and we had group singing. We didn't make it to Prairie View with the singing, but we did go to Prairie View with the math and spelling.

—Ruth Skinner

You can think of so many of life's lessons that those two teachers reinforced. Talk about parents and teachers being on the same page! . . . I think the two teachers, and I think this is true of so many communities like ours, they are just the hub, the glue that holds the communities together. They'd have PTA meetings and every parent would be there. It wasn't as if you could misbehave and not experience any consequences, because that wasn't going to happen, because even if your parents didn't know, your aunts or uncles would. And I think that's one of the things that we've lost with the diminishment of our community. Our kids don't have that safety net under them, that sense of community.

—DeMorris Skinner Young

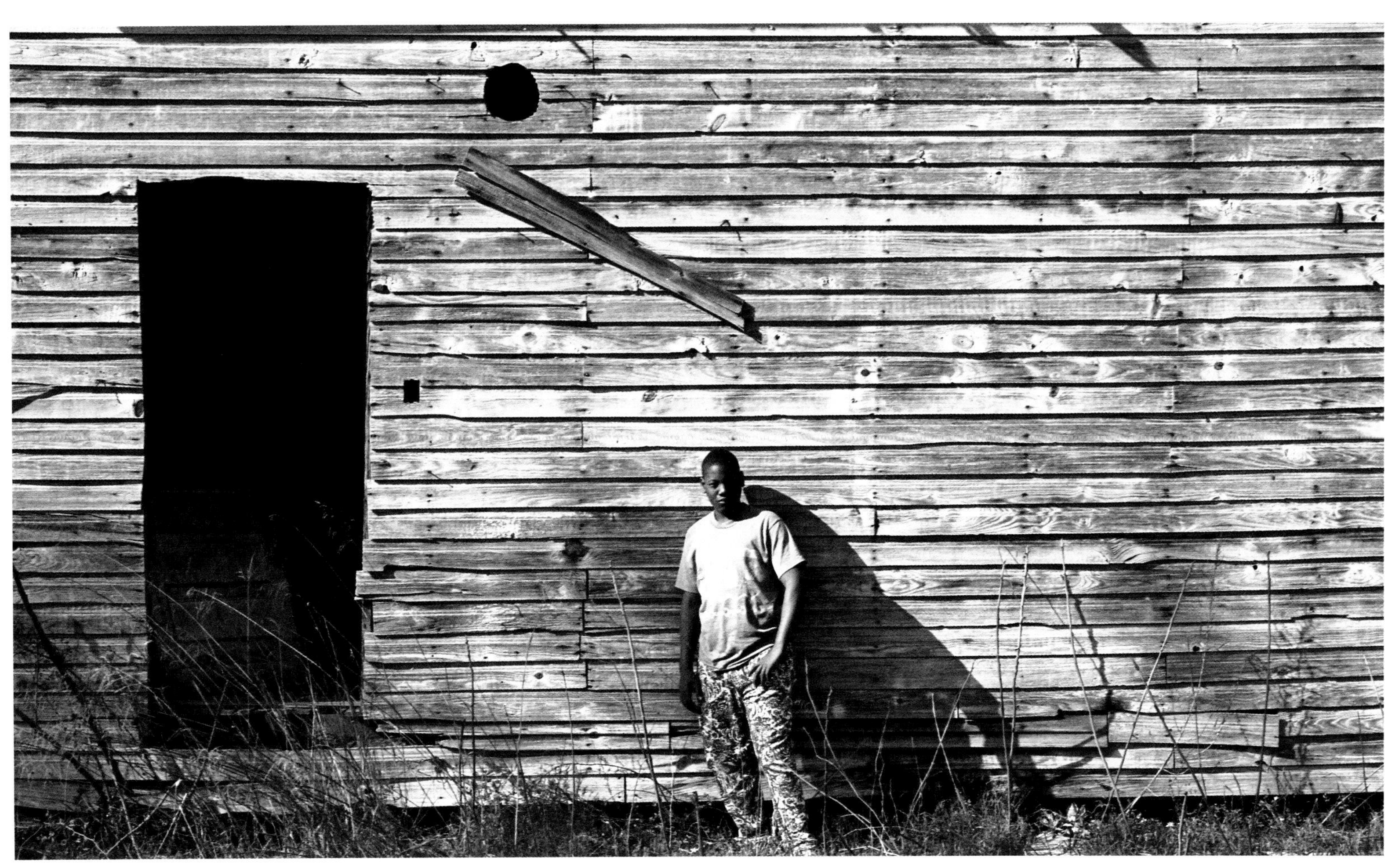

Tearance in Front of the Old School (1993)

We try to drum into their [younger generation] heads the importance of keeping their heritage because of the struggles that our forefathers and foremothers endured and experienced in order to have a little piece of land.

—Maye Upshaw Ham

I think that knowing about this place is the beginning. What will bring them back? I don't know . . . County Line may not be County Line as we know it in 60 or 70 years, but we are exposing our children and our grandchildren to it. And if they decide that there is nothing better out there, then they can come back. That is our hope. There may be something better out there for them. But to us there is nothing better. The best we can do is make them familiar with it.

—Faye Upshaw Johnson

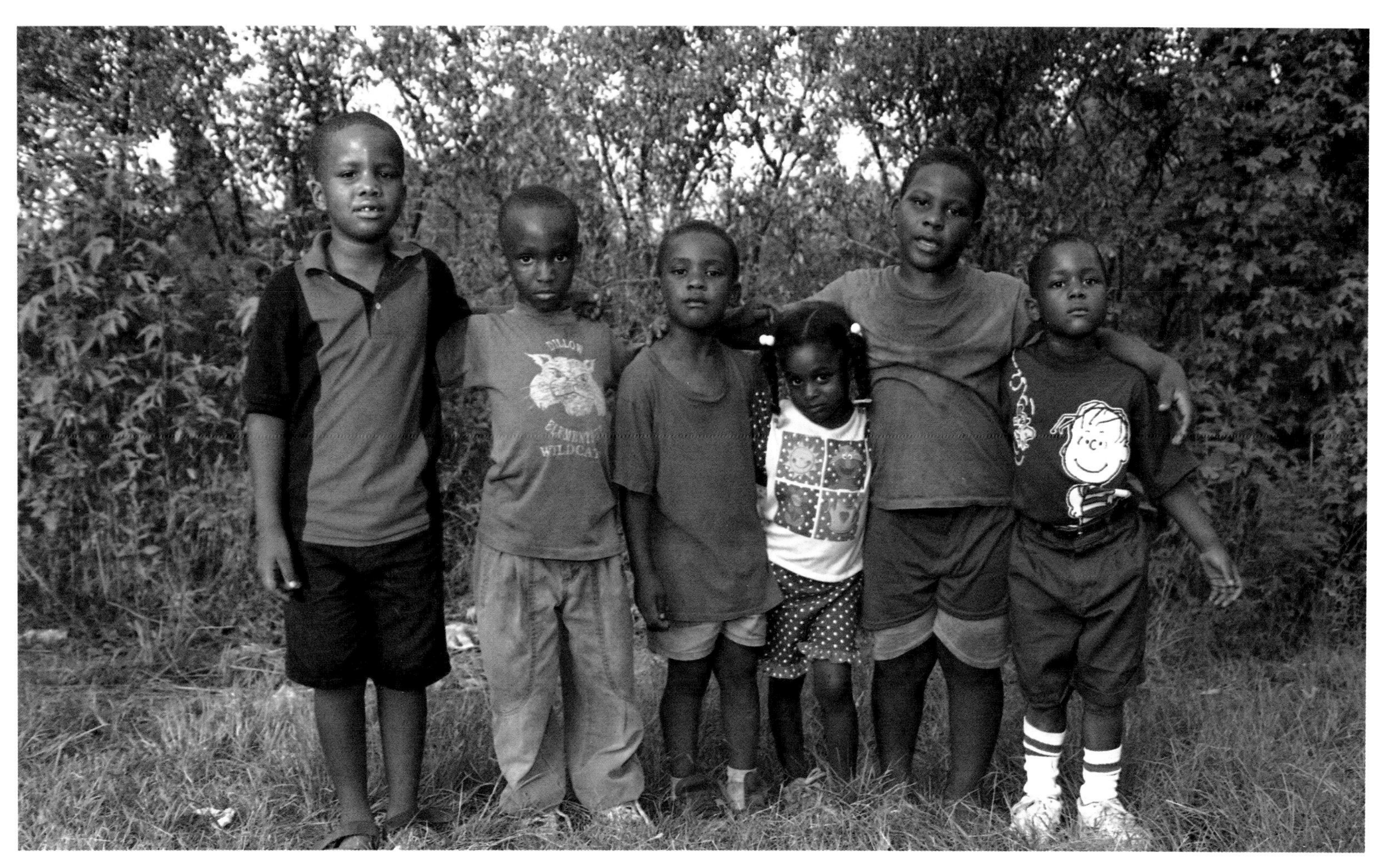

Daquin, Donte, Alfred, Sherrell, P.J., and Tad (1994)

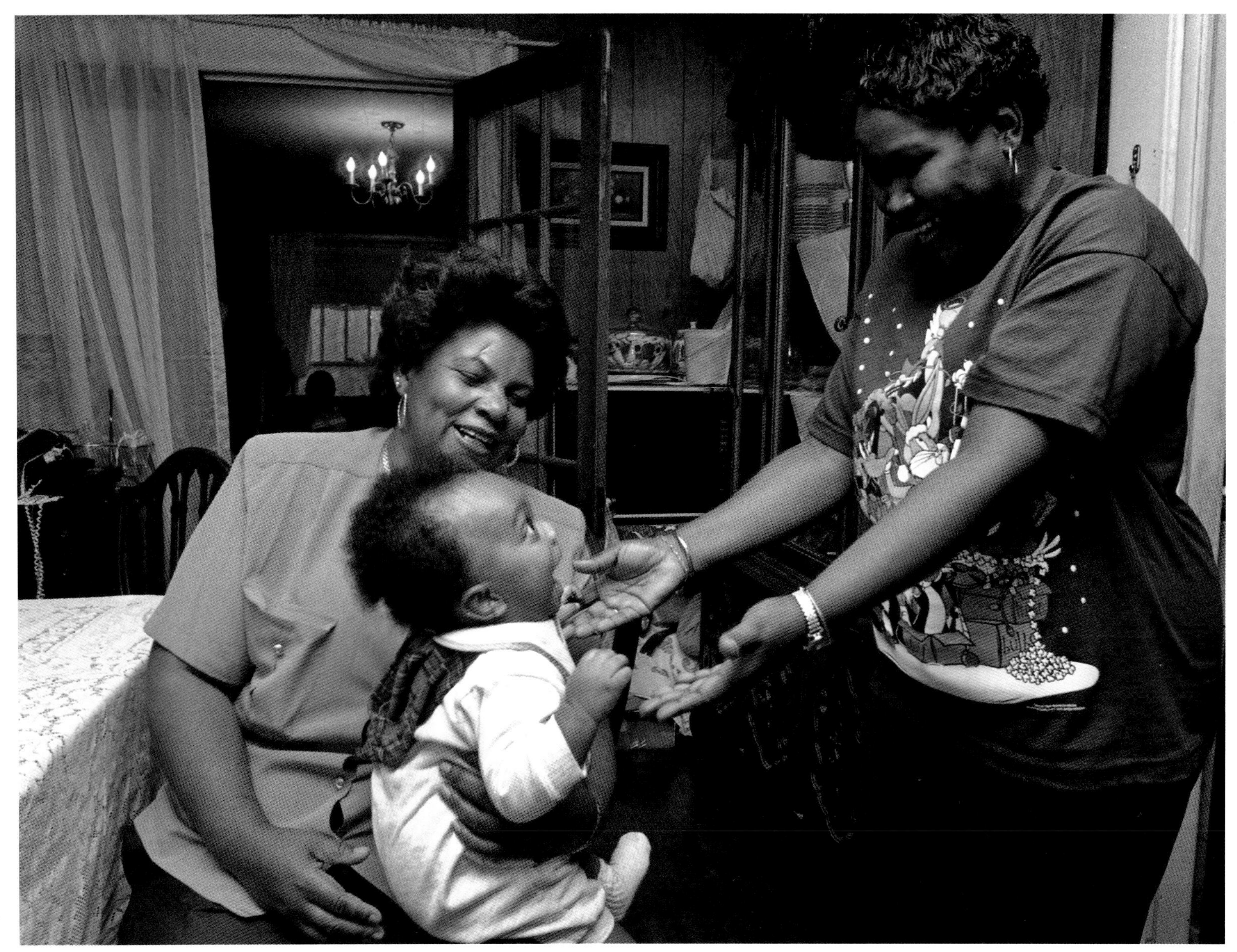

Timothy's First Christmas (1996)

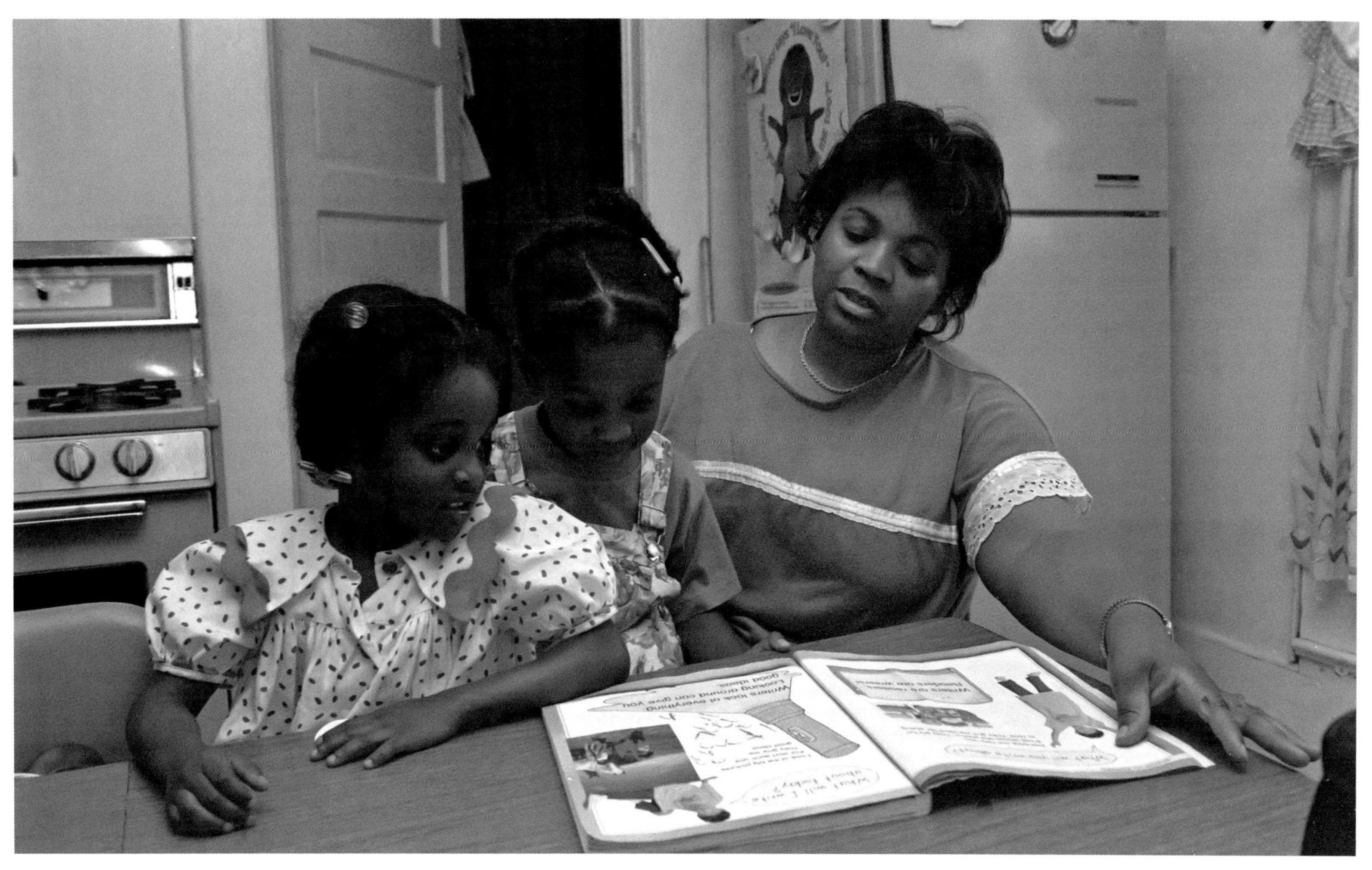

Sherrell, Elia, and Faye (1994)

It was a loving community. We didn't know what it was to have money. We were a community of sharing people. We all were reared by the matriarchs and the patriarchs of this community. We didn't know that we were poor. We raised everything we ate except for the sugar and flour. We were taught to be self-sufficient. My mother canned 500 jars or cans of food every year because she was feeding 18 people, and also we shared. The only time my dad had any cash money was when he sold the cotton crop each year. Meanwhile, we lived on credit. I didn't realize I was poor until I moved away from here. This was a wonderful community, and still is. That's why I came back.

—Dolvin Upshaw

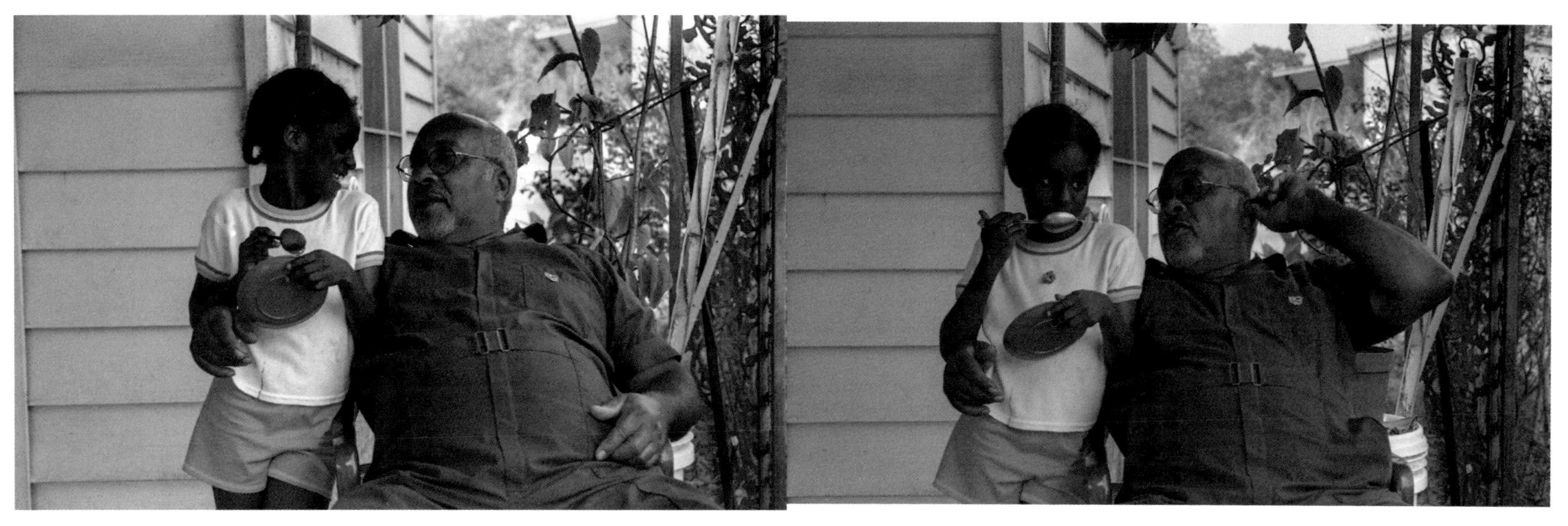

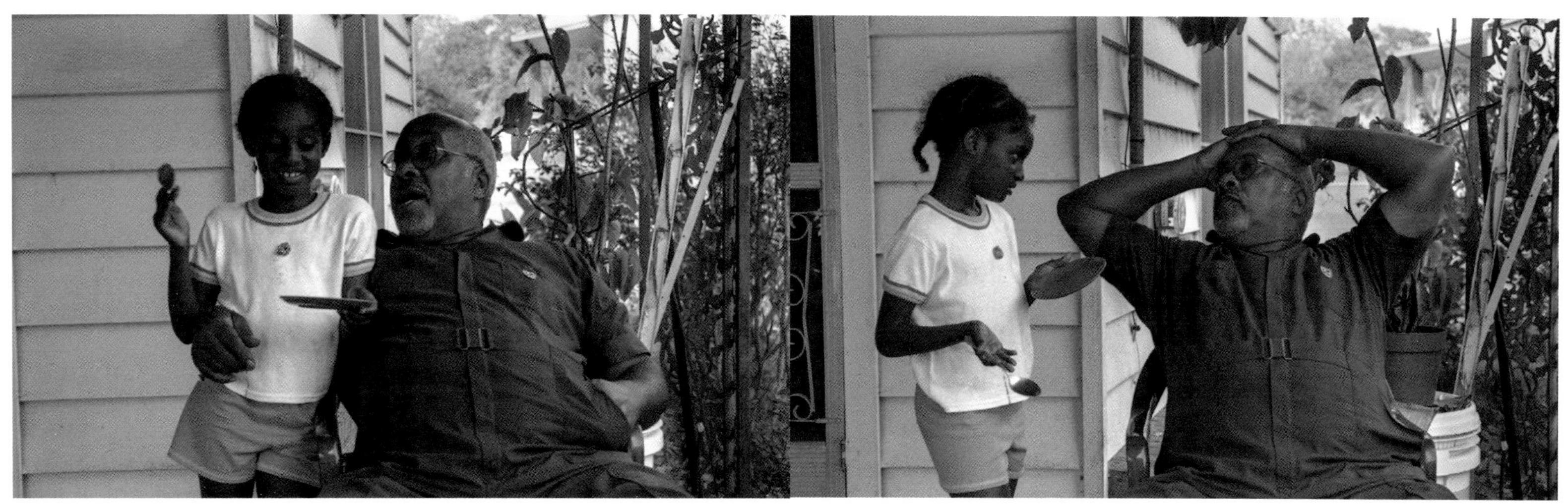

Sherrell and Dolvin in Discussion (1997)

I can tell you what I think our goal is for County Line. And that is to have more of our offspring come back. And build big and better homes. And to declare their land, and to improve their land. And that's why we bring our children and our grandchildren and have them participate in these Homecomings. We don't have to teach them to love the community, County Line. That's natural. They see how much we love it, and I can only speak for my children and my grandchildren, they love County Line. And the future, having been exposed to the rest of the world, I think when we incorporate our children into the community there can be an improvement in the roads and in the landscape without demolishing the natural environment. I think, and it is my hope, that our offspring will want to stay in the community and make improvements and, like I said, declare their heritage.

—Faye Upshaw Johnson

And a lot of my grands, this is what they want to do. They are talking about it. They talk about it all the time.

—Ruth Skinner

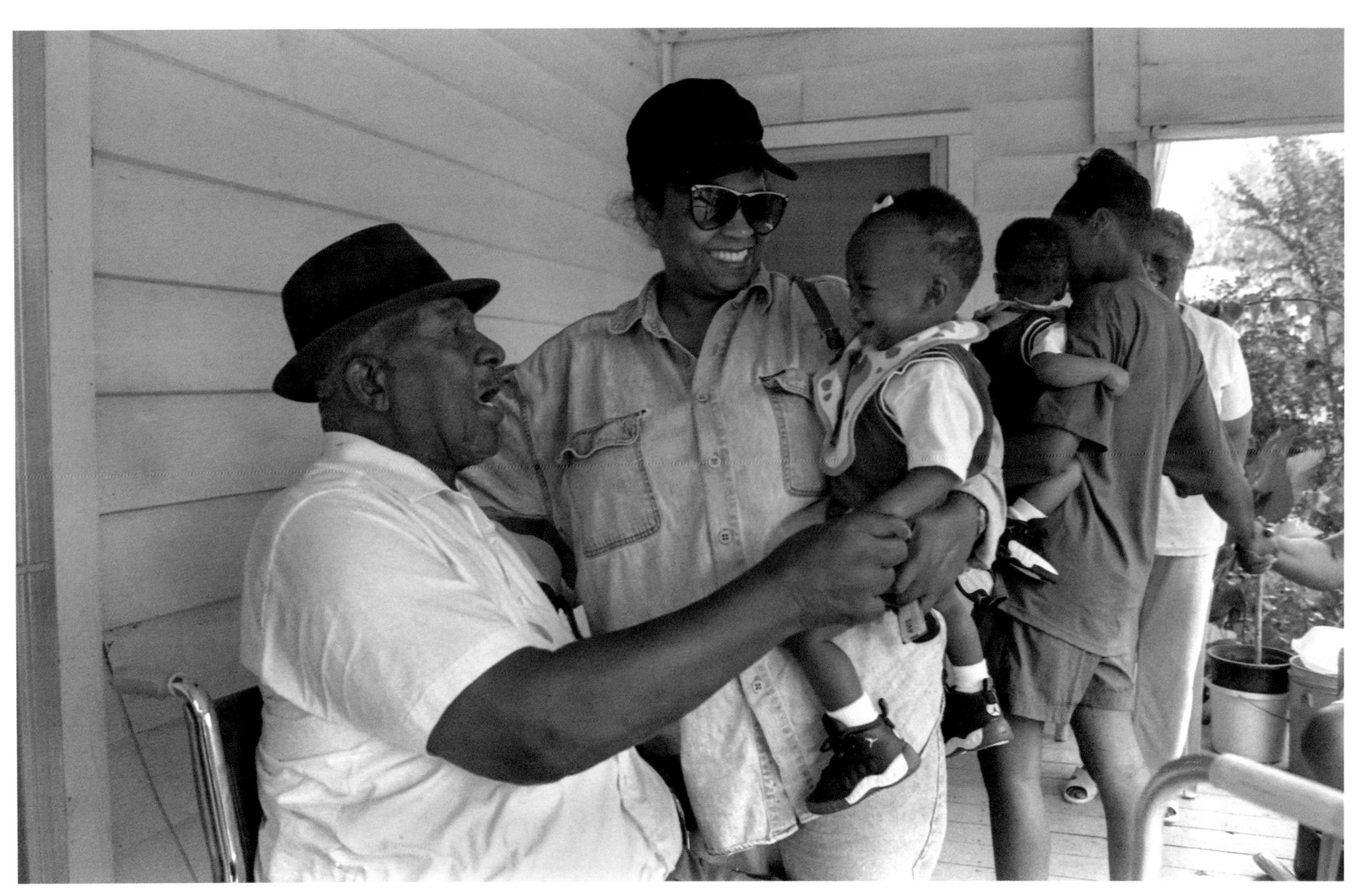

Faye Introduces Reya to Monel (1997)

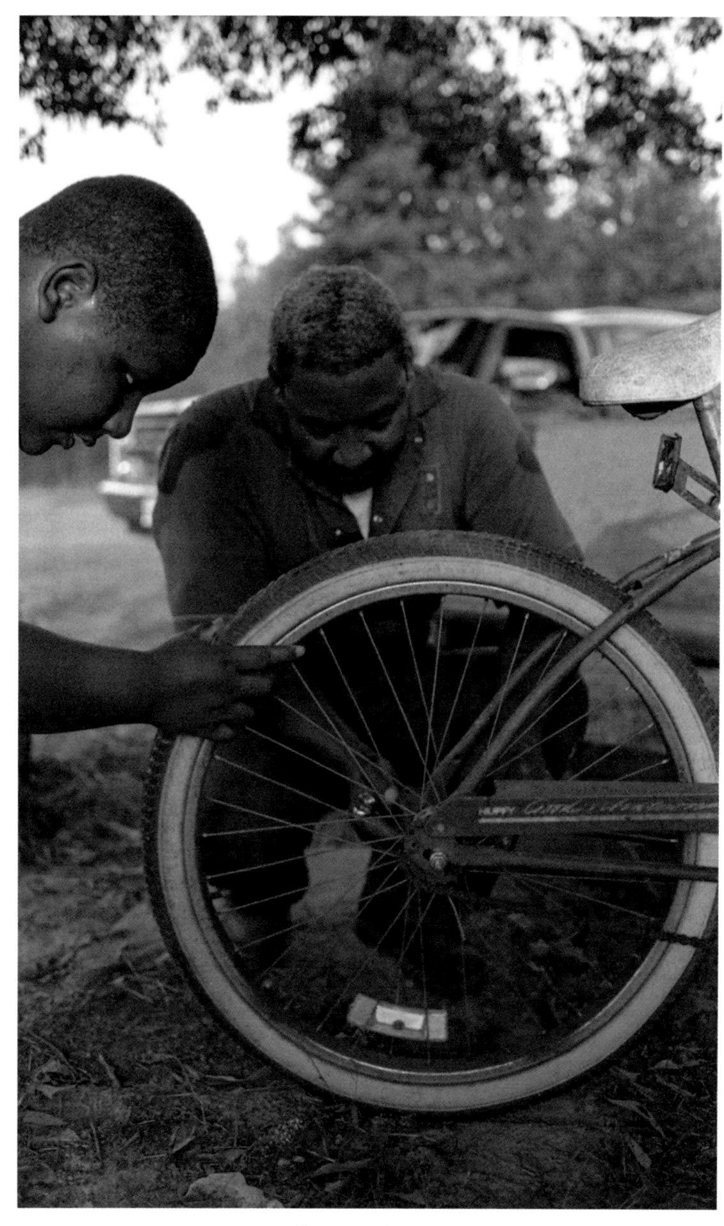

P.J. and Tenchie (1999)

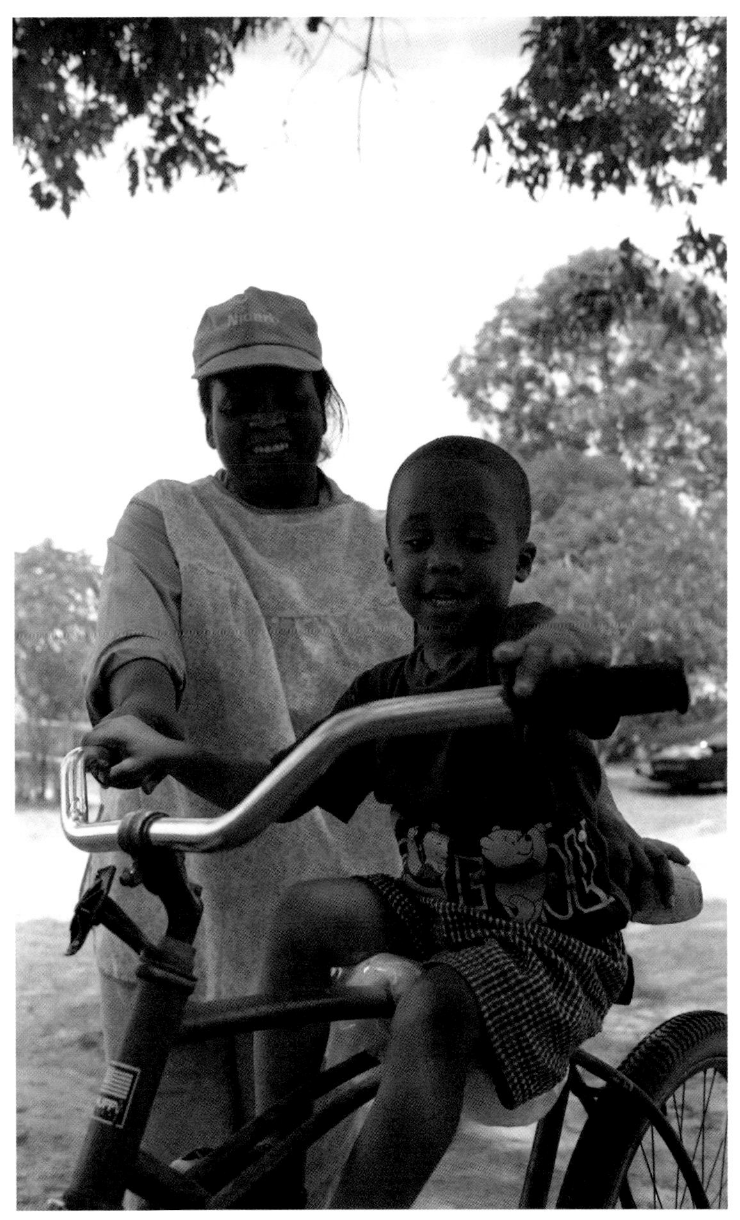

Faye and Rashad (1999)

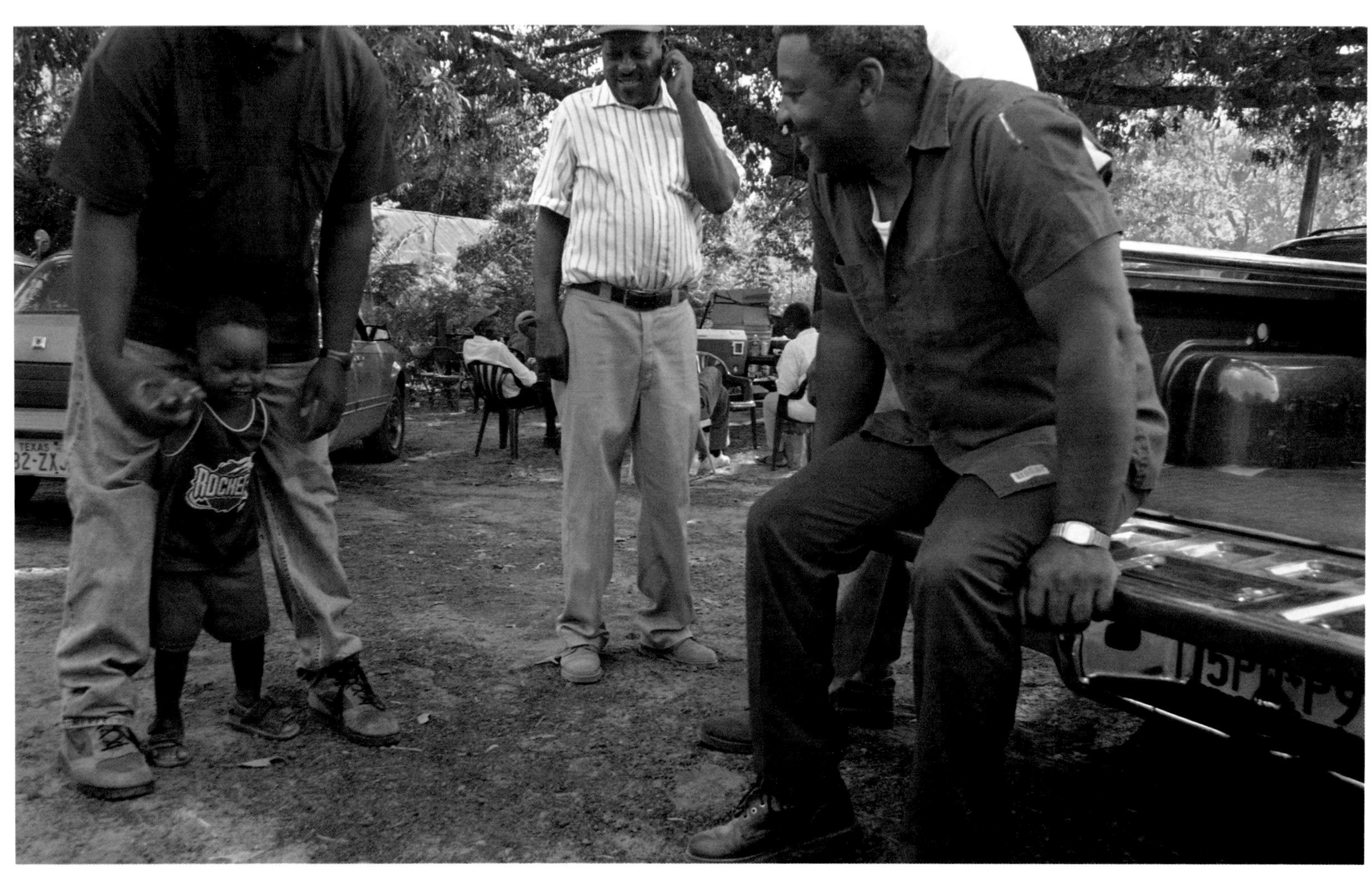

Brandon Learning to Walk (2001)

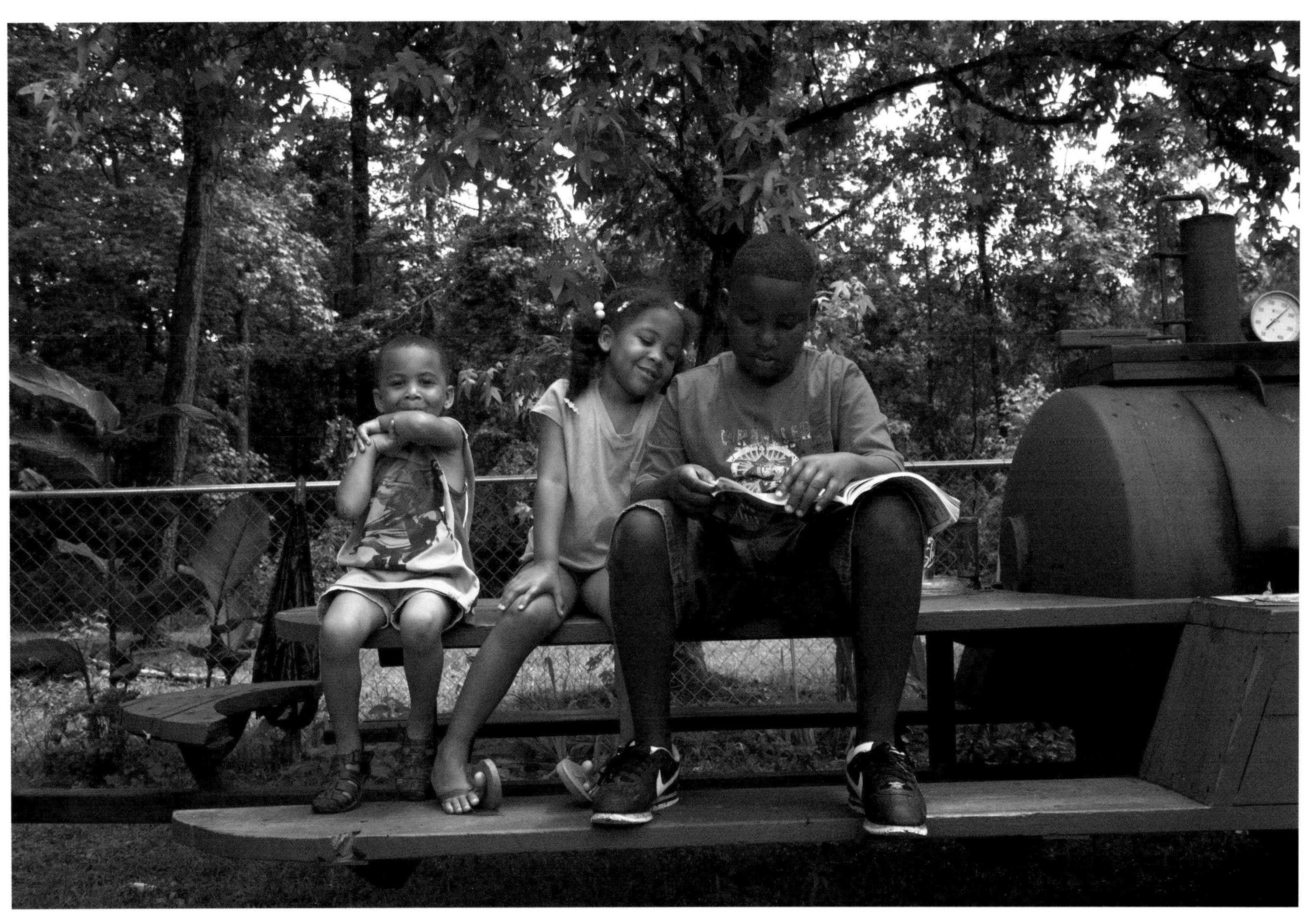

Trent, Teralyn, and Brandon (2009)

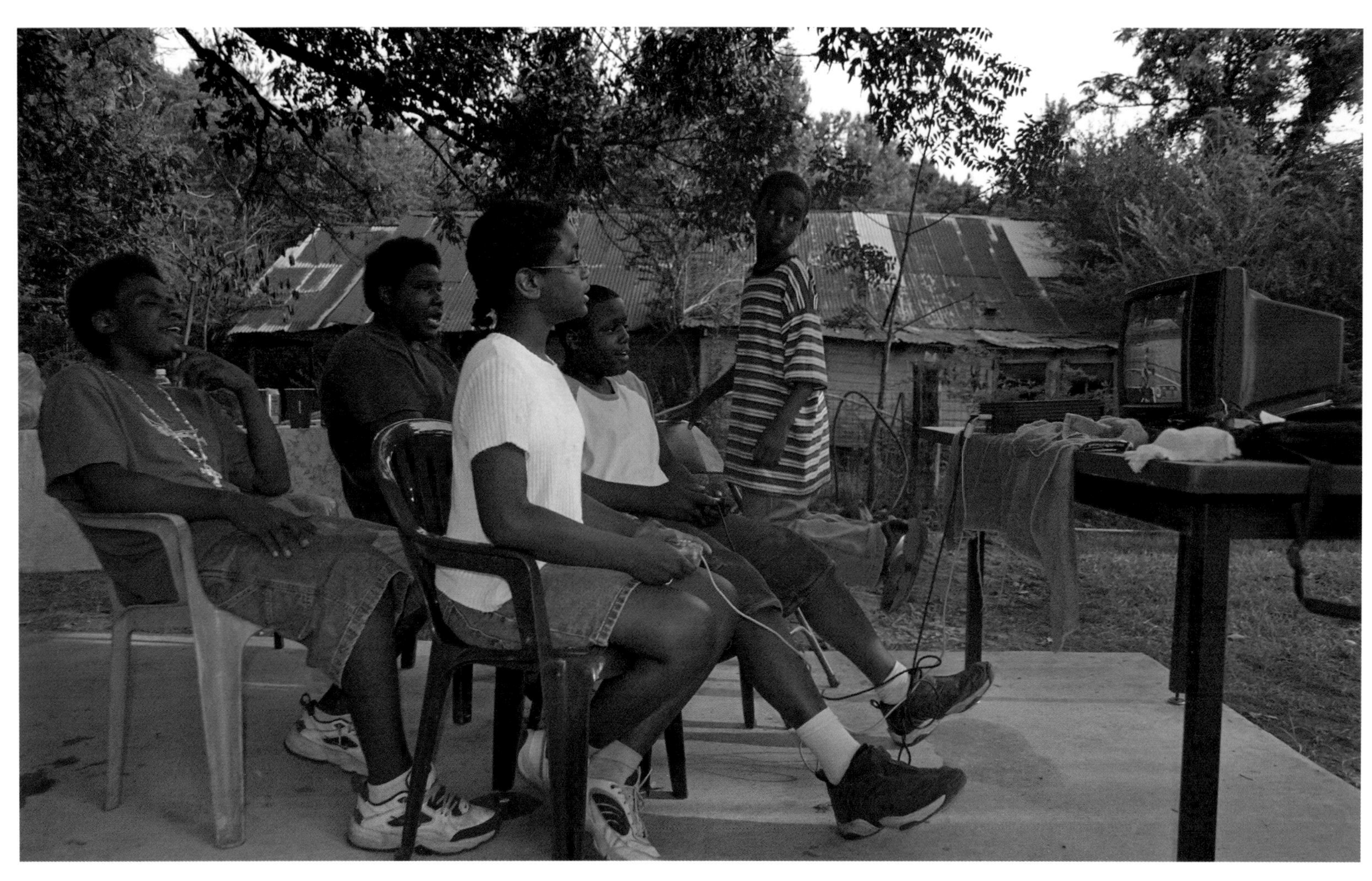

Alfred, P.J., Elia, and Tad Play a Video Game at Homecoming (2002)

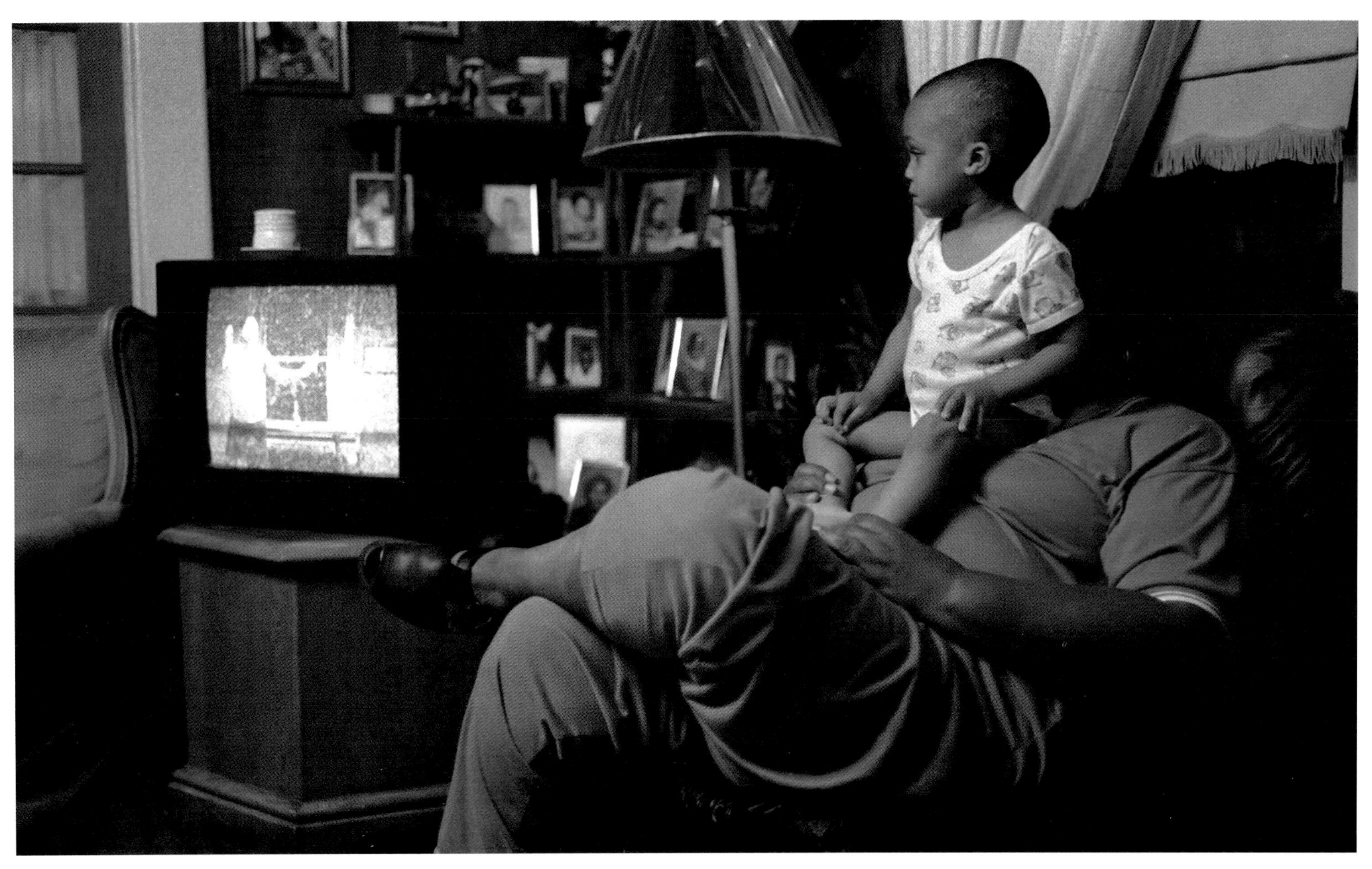

Rashad and Marceline (1998)

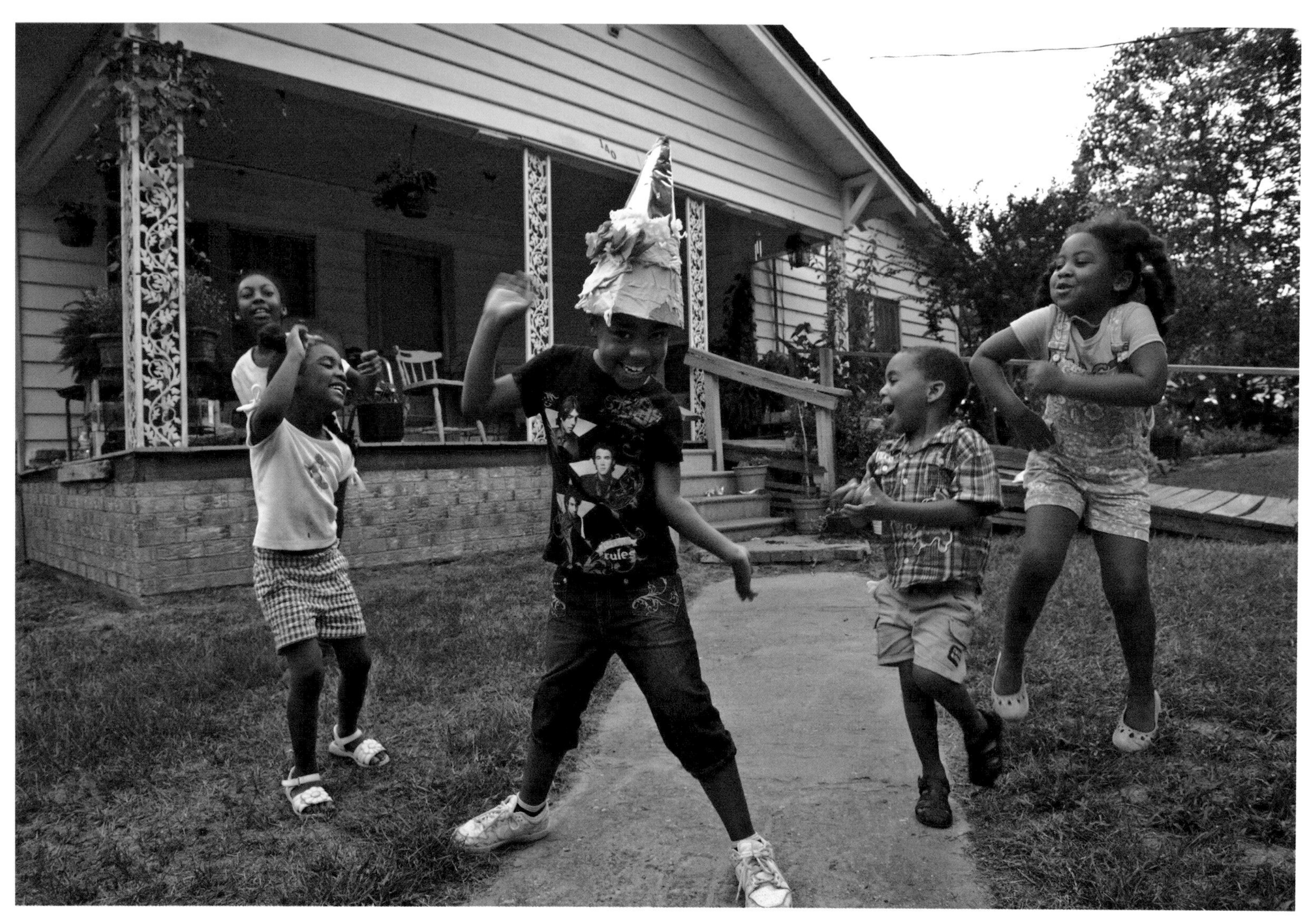

Kiah, Joy, Sky, Trent, and Teralyn (2009)

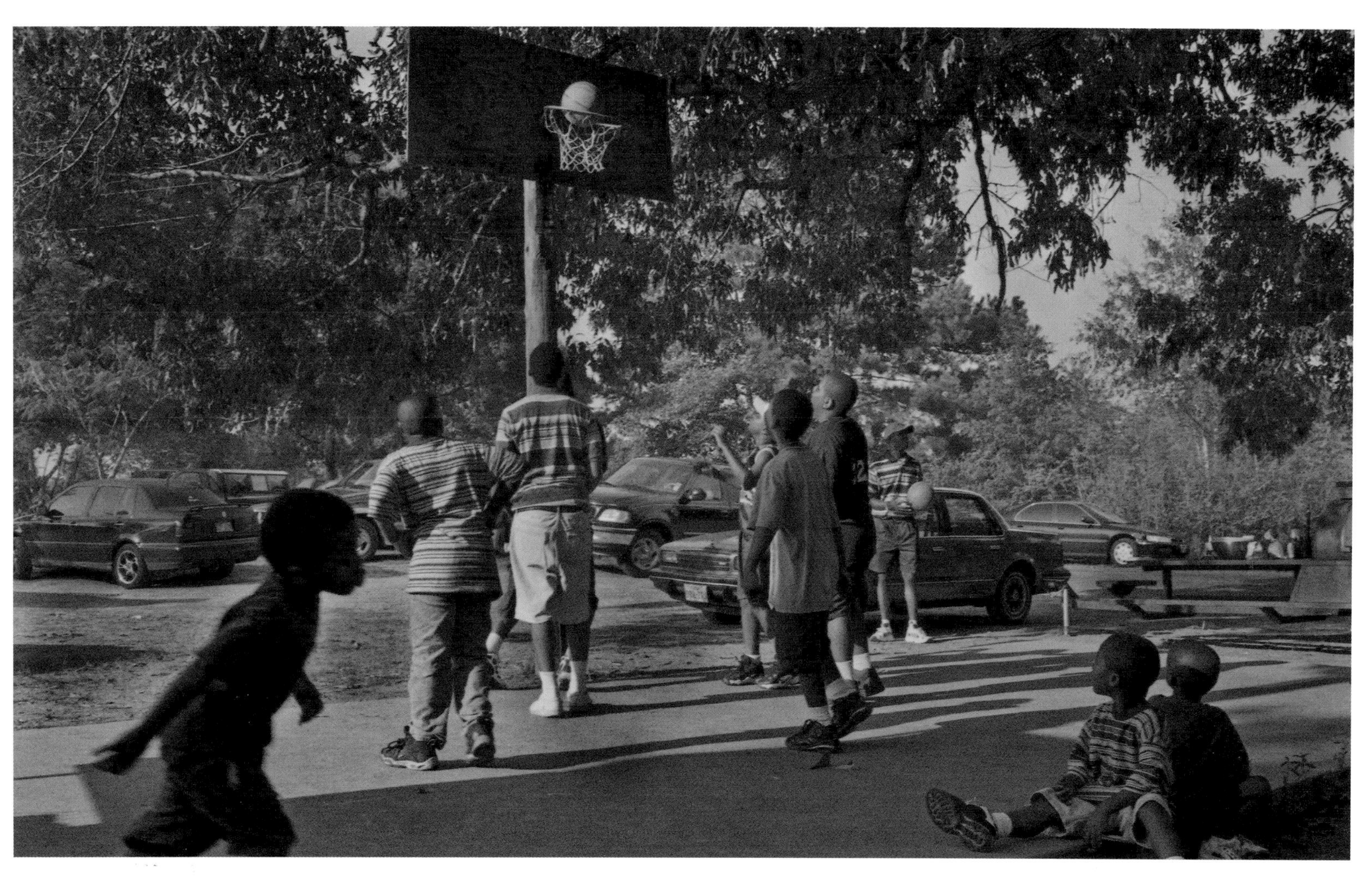

Children Playing during Homecoming (1998)

Saturday mornings or more often Friday evenings heralded the beginning of the young ladies' beautification ritual. First, there was the hair washing and the drying. No one in the community had a hair dryer, not even the beautician, our Aunt Odessa. It took all night for our hair to dry. Several hours in the sunshine would accomplish the task as well.

Once the hair was clean and towel-dried Mama or an older sister undertook the massive task of combing out the naps. Naps are notoriously stubborn tangles. Combing through them was accomplished amid facial grimaces and occasional yelps. The tools chosen for this task were of vital importance. The wider the spaces between the teeth of the comb the easier it was to tug through the hair. Combs with narrowly spaced teeth were shunned, hated tools of torture. They were the source of many a sore head.

When finished, our heads were adorned with at least a dozen small plaits. The smaller the plaits, the quicker the hair dried. Even though such small plaits were a nuisance and certainly unattractive, they were necessary because large sections of hair took much longer to dry.

—Beatrice Upshaw

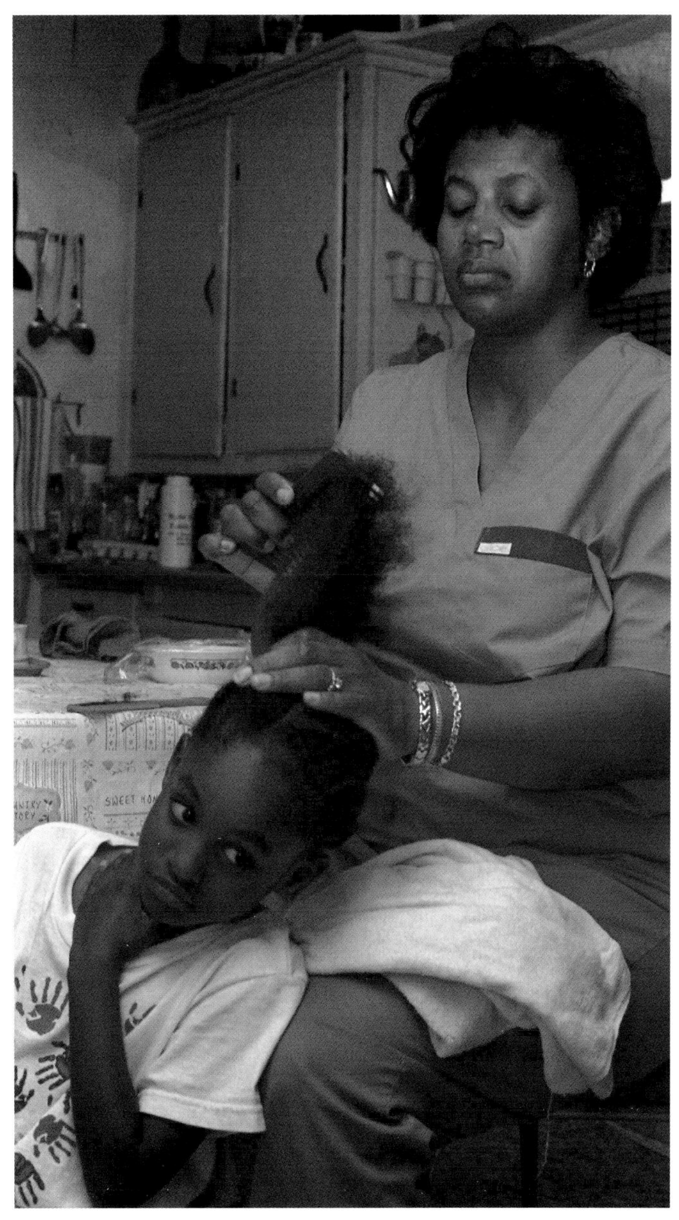

Sherrell and Beatrice (1997)

The first hog that each man would kill in the fall was shared with the community. All the brothers would be there and help dress the hog . . . chewing tobacco and telling lies. They'd hang that hog up, clean it, all pitch in, and when they'd get through, "Brother, what part you want?" That hog was shared with every man there.

—Dolvin Upshaw

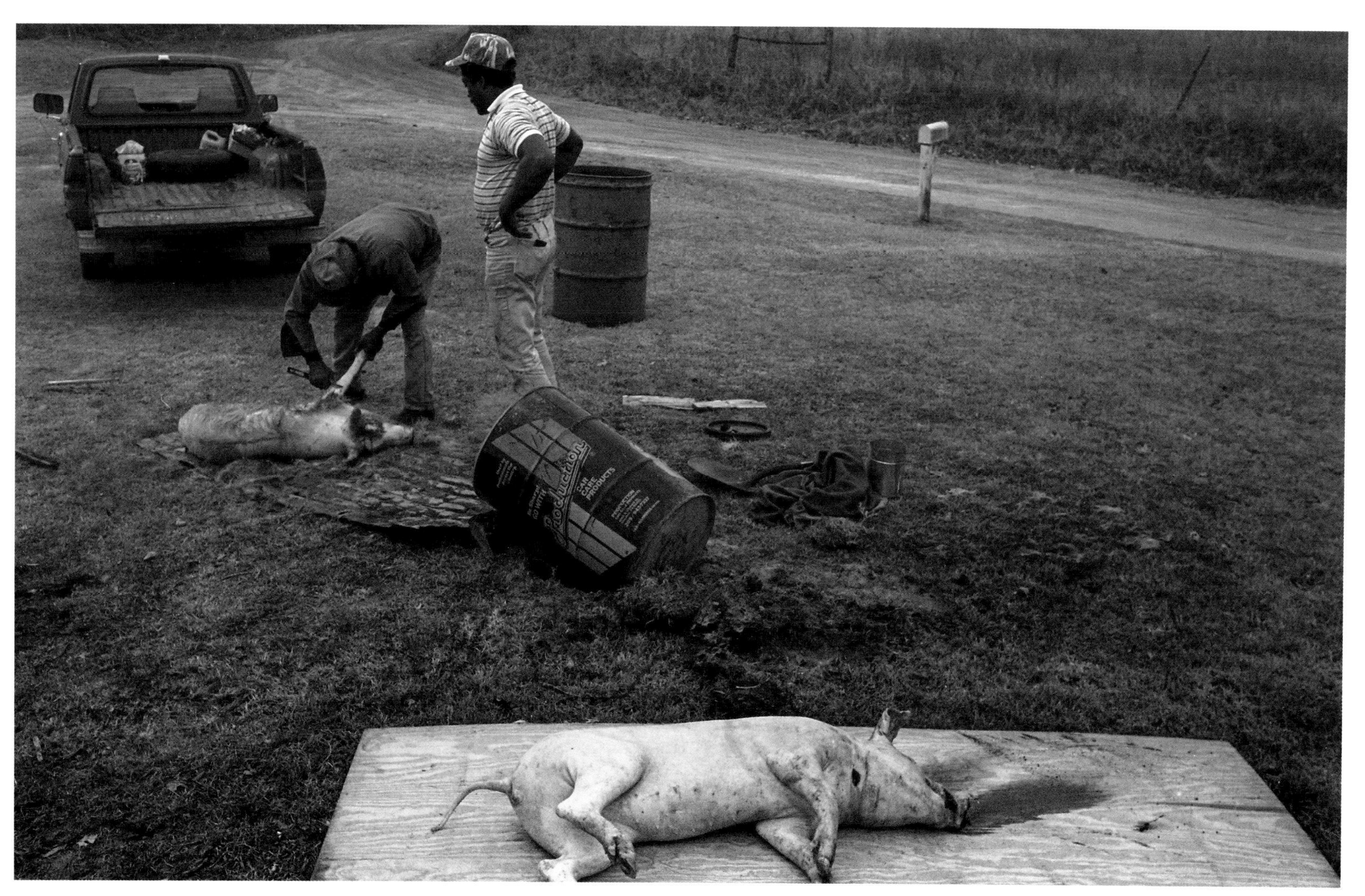

Hog Butchering (1990)

If you really want to know the community's gathering place, it was the store. That's where all the lies were told, and all the fears and aspirations came out. That's where they'd hoorah each other and sometimes get angry. But it didn't mean anything.

—Claude "Bubba" Upshaw

The general store was by far the most popular hangout in our small close-knit rural community. Well, really, it was the only hangout unless you counted the church across the road. It was not to the church that we turned for basic entertainment though. Rather, the store was the place to be on Saturday. Community residents hung out at the store morning, noon and night. My paternal uncle, Claude Upshaw, owned and operated the business. Inside, there was an ancient black and white RCA television, surely the prototype, and an even older brown plastic tabletop Philco radio. That's it. There was no jukebox, no pool table, no dance floor, and no record player. There was nothing to draw people to the store except amiable fellowship and innocuous amusement . . . The opportunity for socialization was the real motivation for our frequent trips to the store.

—Beatrice Upshaw

Our Dad said he'd rather play dominoes than eat. He'd go up to the store and play. He said he couldn't be beat. People would come down to the store and play with him. He'd say, "I'm the kang [king]!" The more trash they talk, the more they like it. There's no money involved so you didn't mind getting beat . . . . I've seen them slap the domino on the table and all of them [the other dominoes on the table] would jump up. They love that game.

—Marceline Upshaw Harris and Claude Upshaw

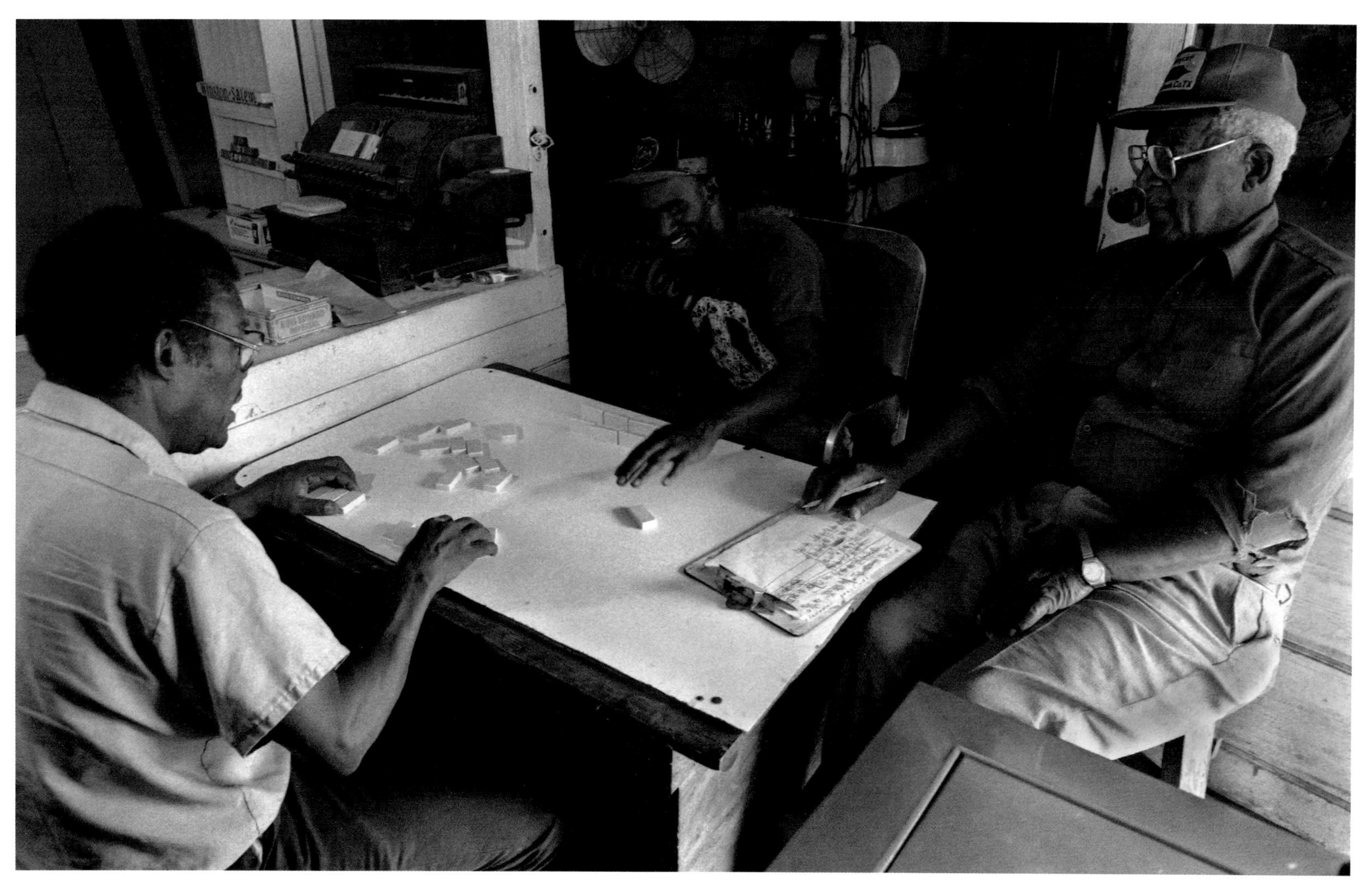

Dominoes in the Old Store (1989)

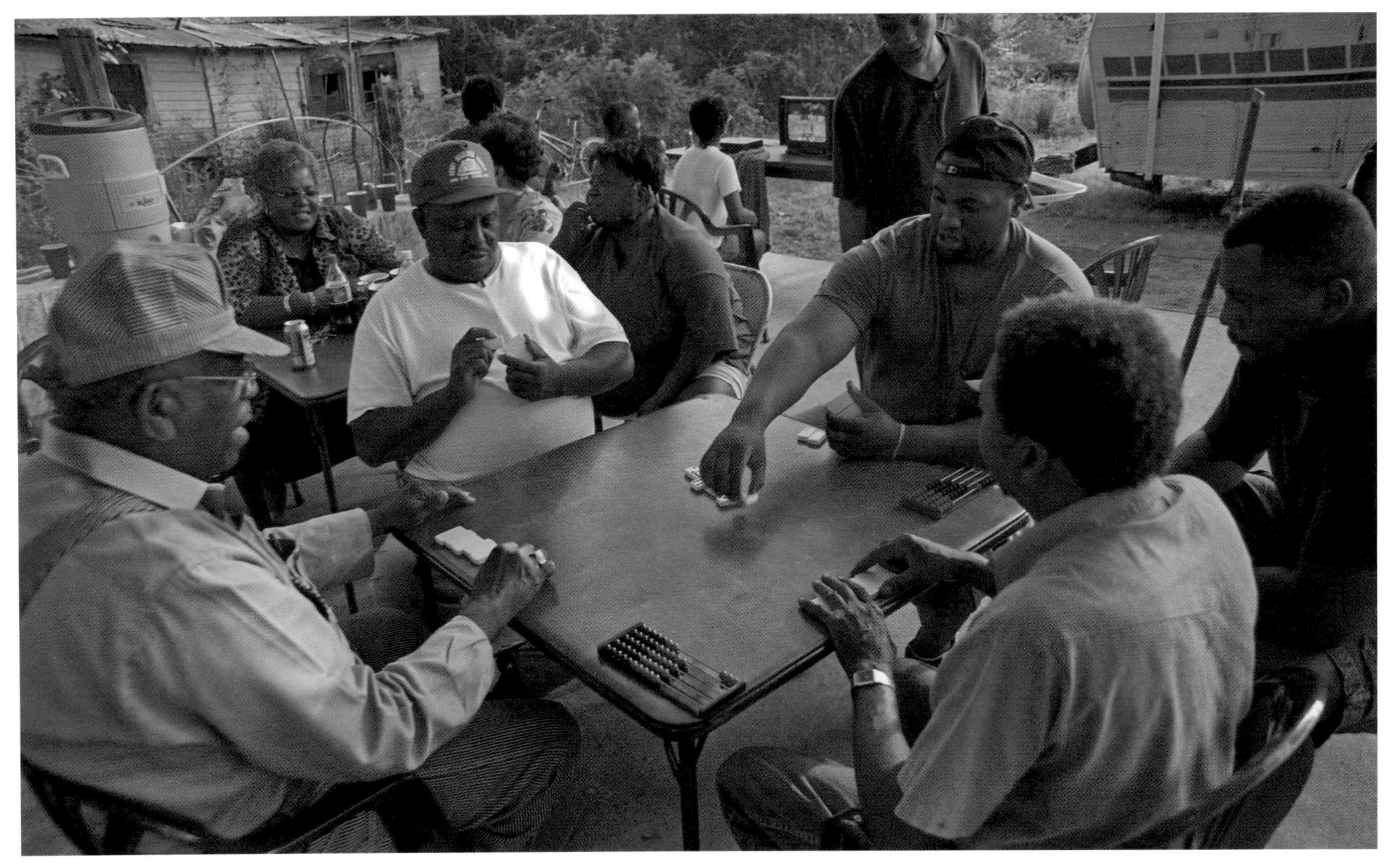

Dominoes (2002)

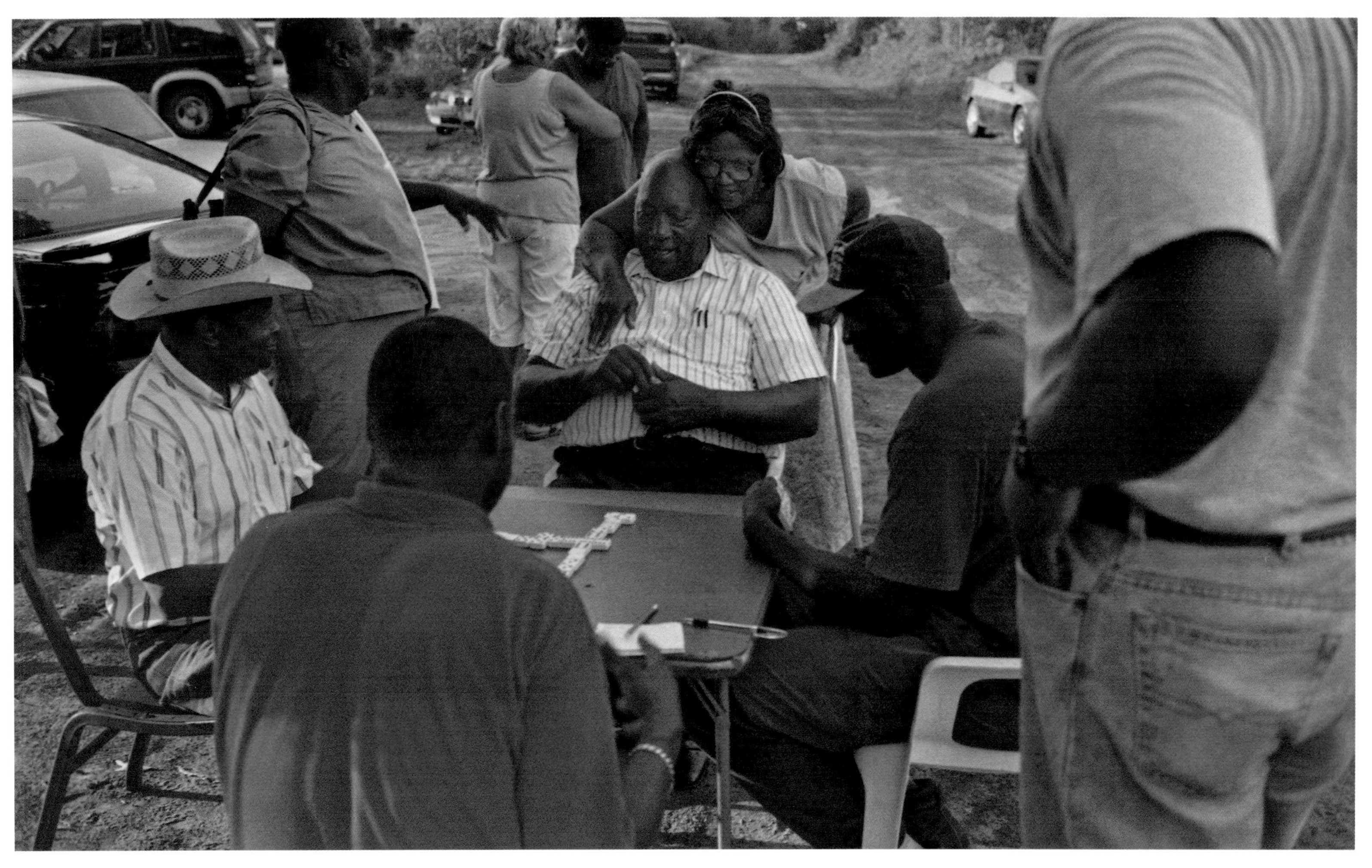

Dominoes (1998)

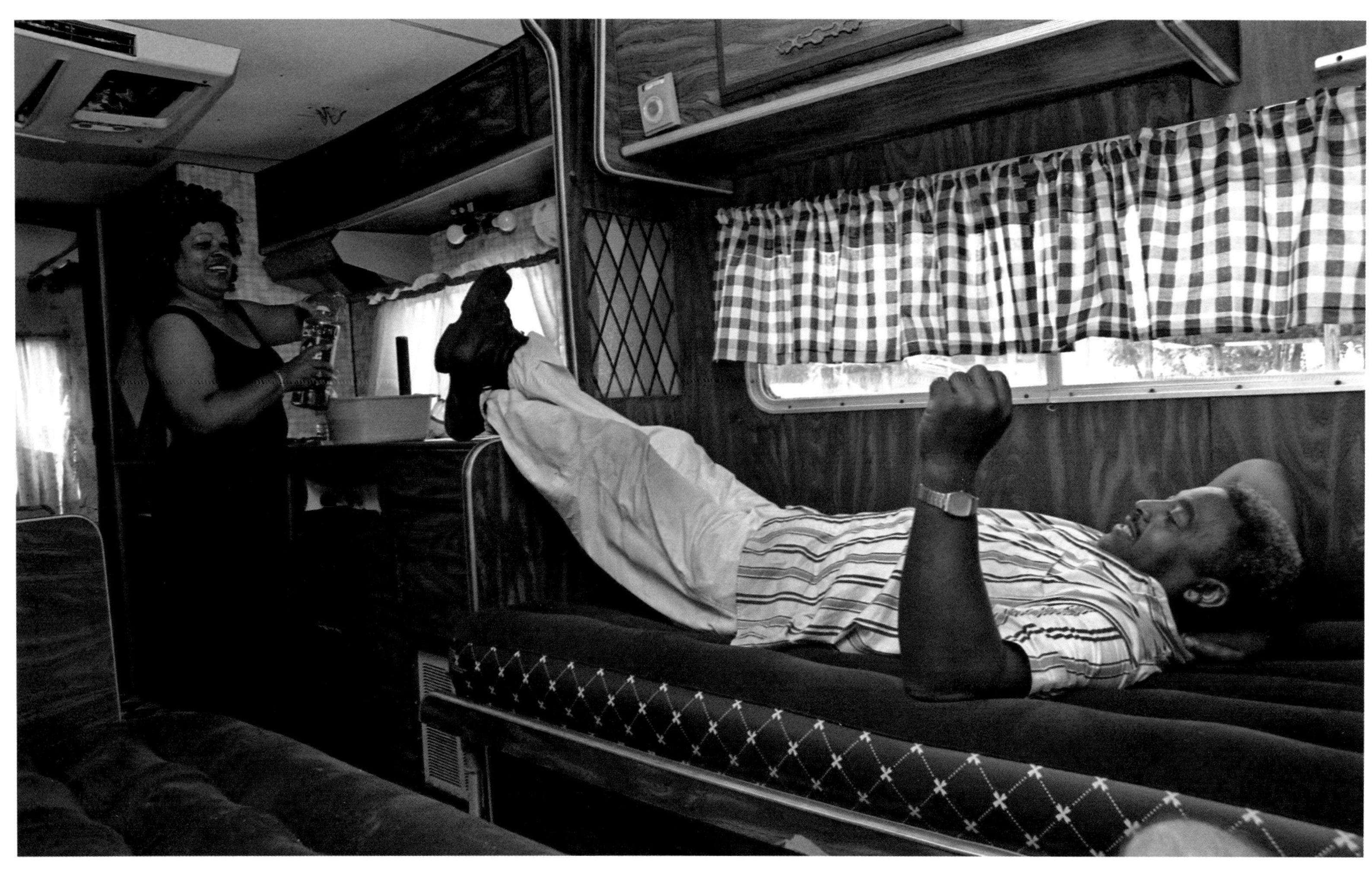

Emma and Tenchie (2000)

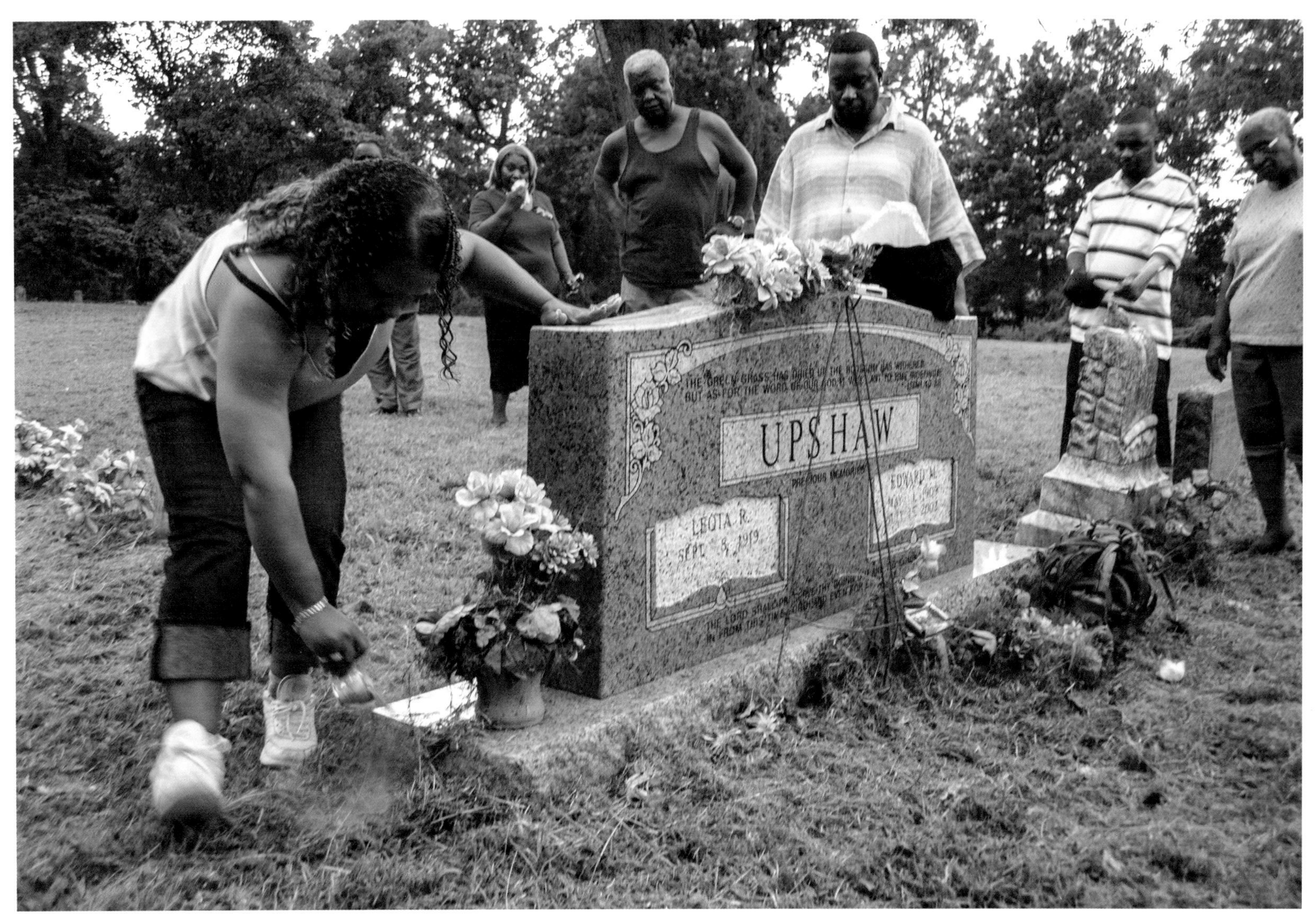

Tenchie's Ashes (2006)

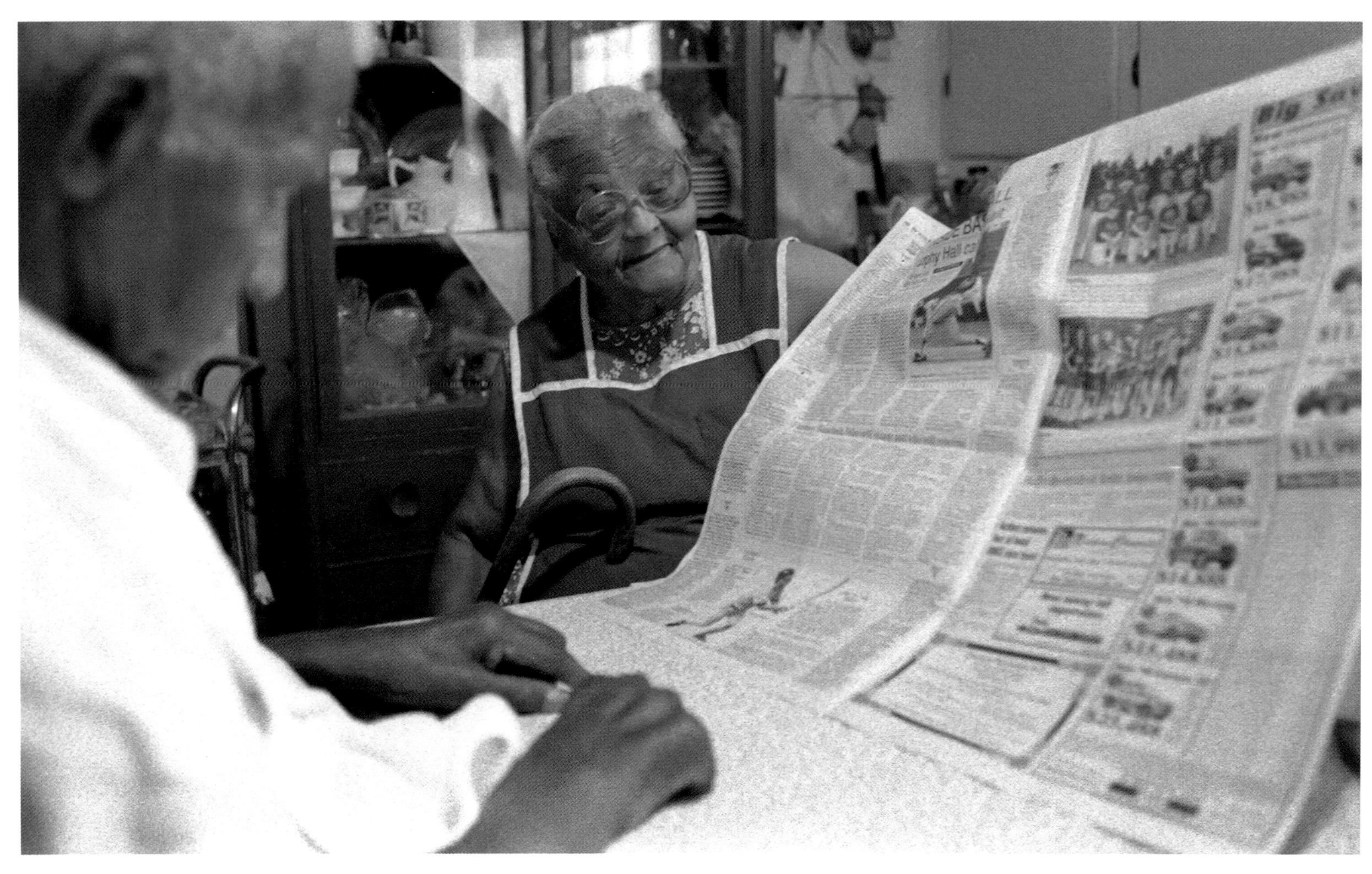

Monel and Leota (1998)

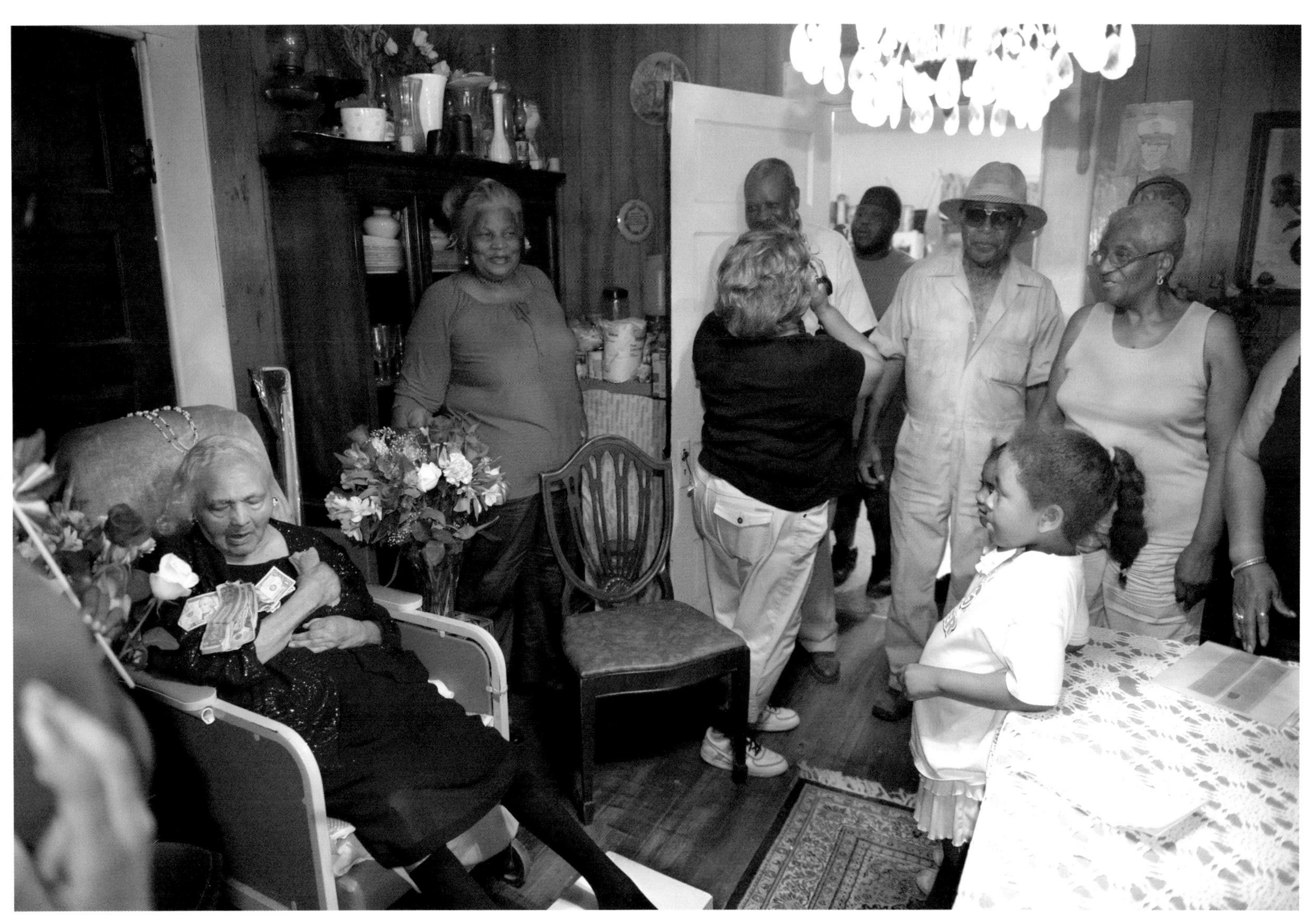

Leota's Birthday (2009)

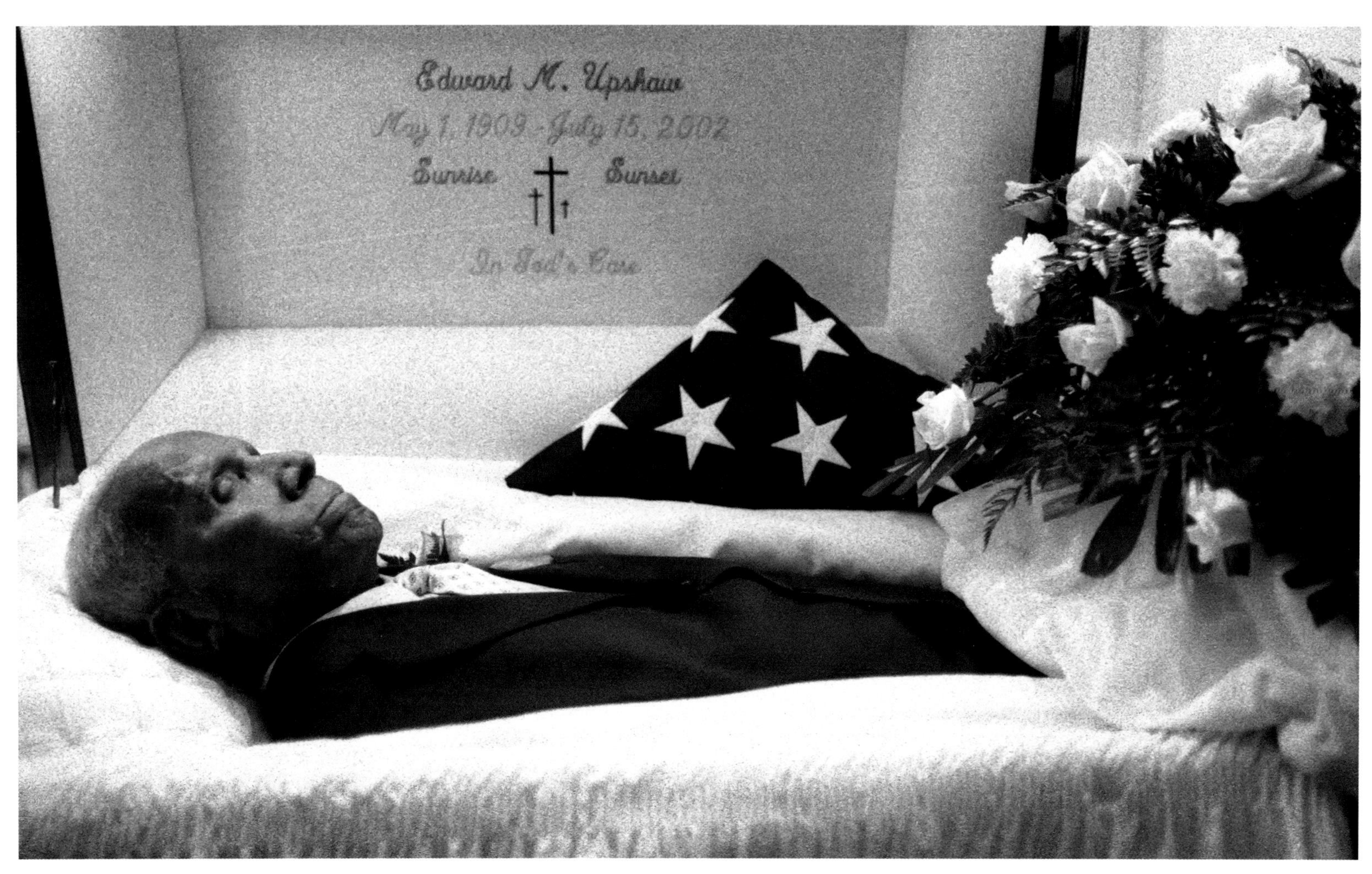

Monel Upshaw (1909–2002)

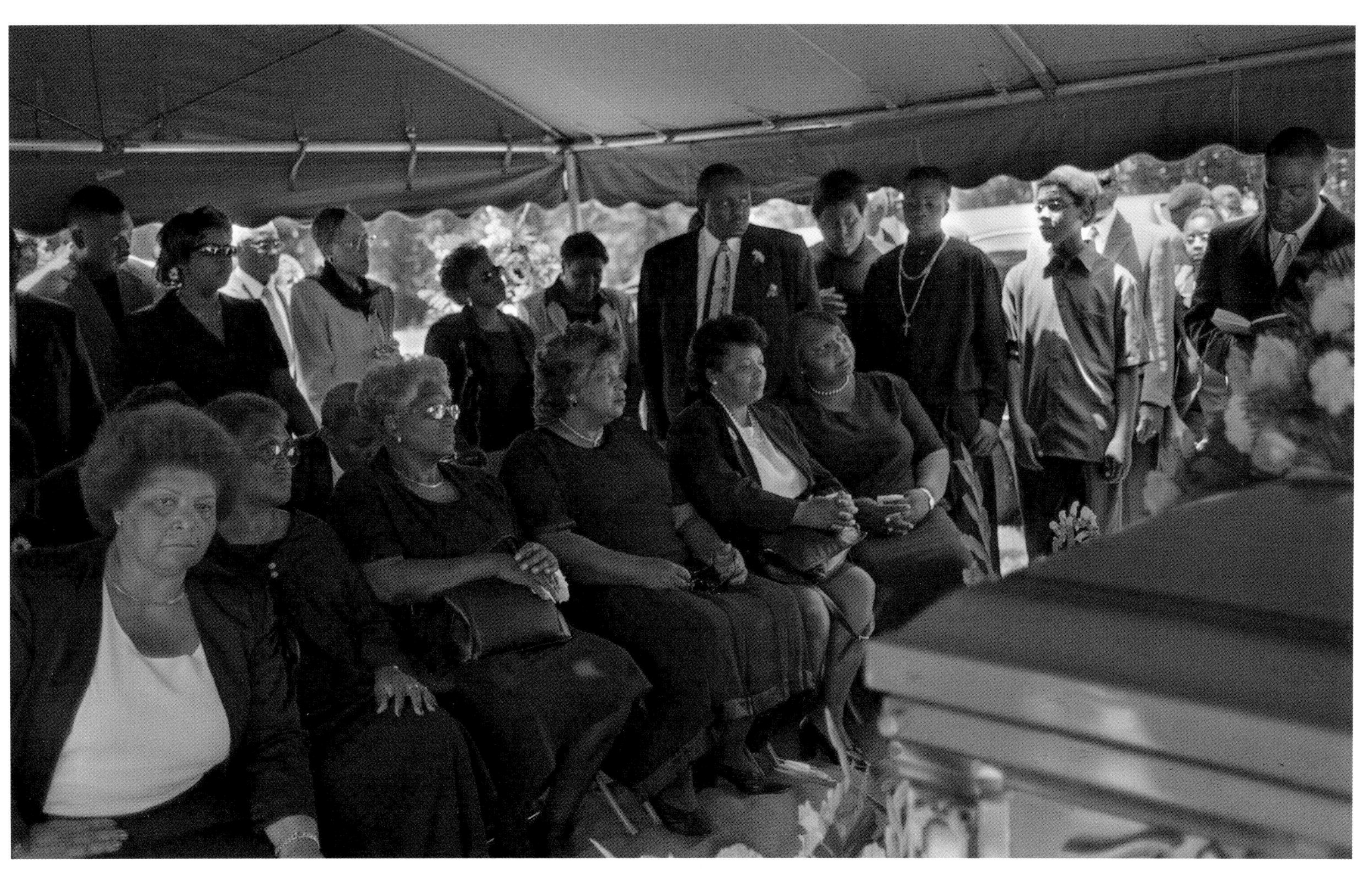

Monel's Funeral (2002)

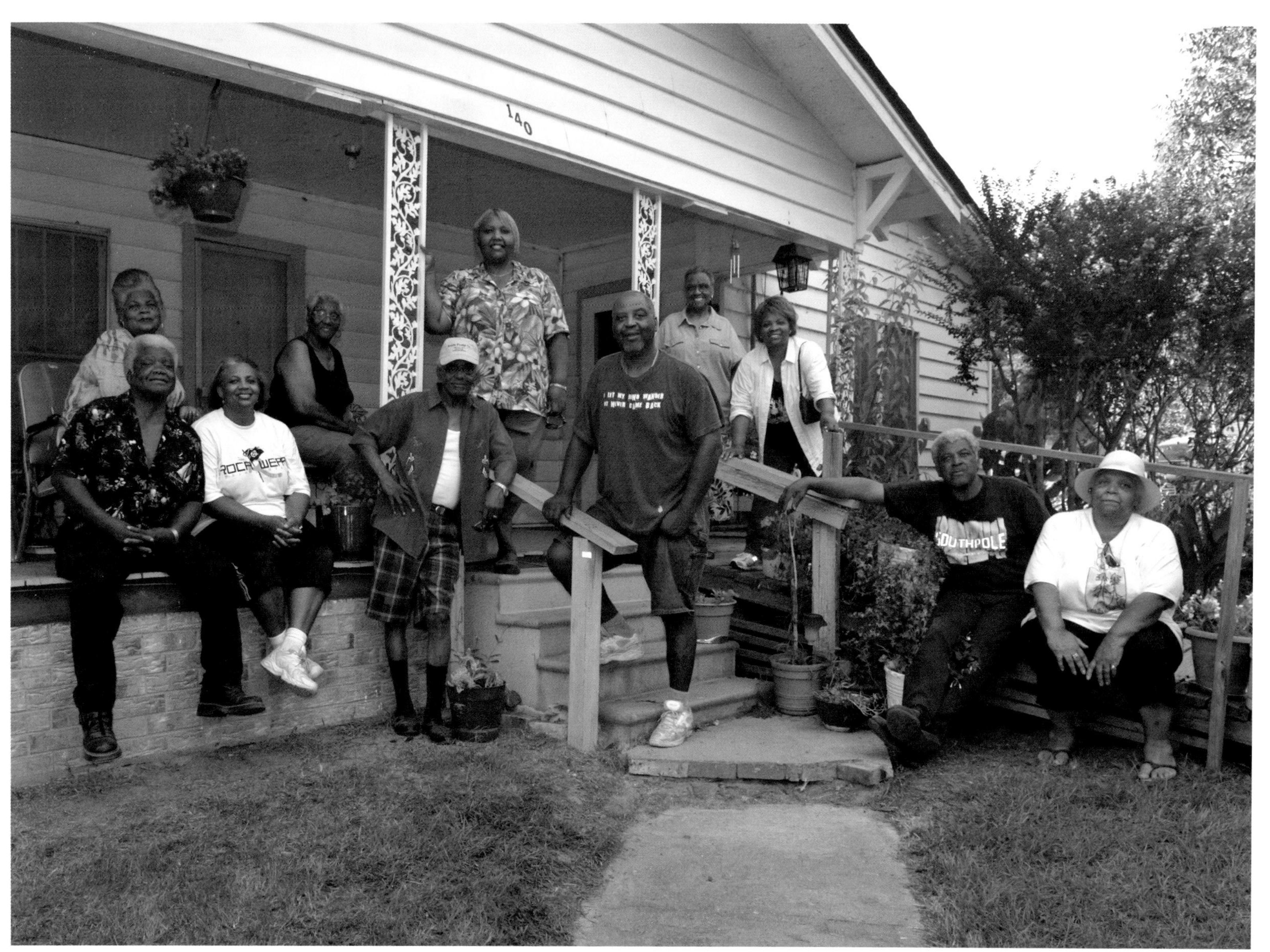

The Remaining Siblings (2009)

Ten years from now there won't be a community because the people are getting older, and those who are not getting older are leaving. And those who are getting older but not leaving will eventually go to eternity. So I don't see County Line being a community in ten years.

—Marion Upshaw

Somebody's always going to be in County Line because there will always be somebody moving back home like there is now.

—Marceline Upshaw Harris

I'm more skeptical. Where there is no youth, the community dies. Everybody goes to Dallas or Houston and gets indoctrinated. My kids don't know anything about County Line. All they got is a memory [from childhood visits] and they don't seem to care about coming back. That's hurtful, but I'm just looking at it practically. It seems like all the youth . . . they aren't concerned about County Line. That bothers me. But, then, her kids...

—Claude Upshaw

[Finishing Claude's sentence above] My kids never went anywhere, just like I didn't. I never wanted to go anywhere. When I leave from here I'll go to the graveyard.

—Marceline Upshaw Harris

Twenty years from now I hope to be back. I don't want it to fade away. I want to keep my land and live there and grow things and have chickens. I don't want my kids to live in a place where their feet don't touch real earth. I don't want to be the person who stays away and gets really old and then comes back. I want to be able to go back and build a house and have a huge orchard and those types of things.

I think when people realize what they really want to do with their lives they are going to want to go back home. Everybody has good memories there. People are always going to want to go back home. So 20 or 30 years from now everything is going to be fine. It will be just like it is now.

—Elia Upshaw Ali

My Favorite Story

In 2006 my foster brother Alfred "Tenchie" Upshaw died unexpectedly and a number of us went to Los Angeles for his memorial service. It was standing room only in the church. A lot of people knew and loved Tenchie. Toward the beginning of the service our sister Marilyn stood in front to introduce the family members who had come from Texas. When she got to me she said, "Our adopted brother, Richard Orton." I was sitting beside sister Faye. Sitting on the other side of her was her granddaughter, Reya, who was then nine years old. Faye later told me that after Reya heard me introduced she had turned to her, looking surprised and amazed, and said, "I didn't know Uncle Richard was adopted!"

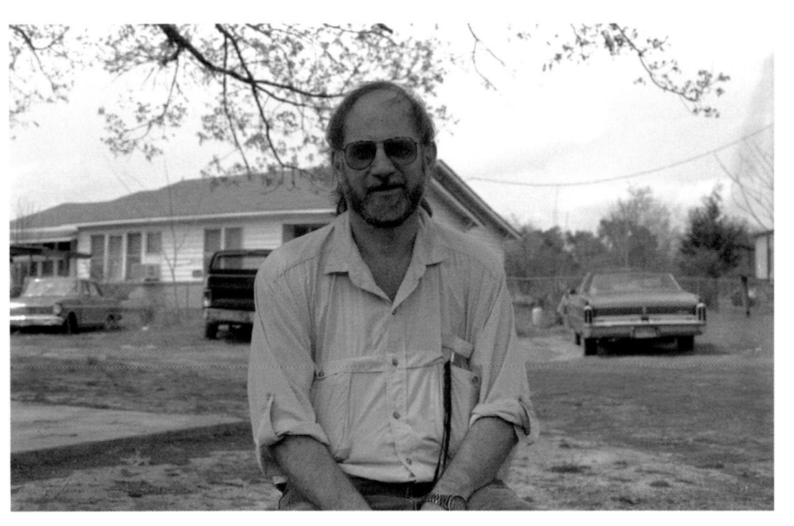

Richard Orton in County Line (1998)